Pirates
of the
Darien

Pirates
of the
Darien

The History of the Espiritu Santo de Cana Gold Mine, Darien, Panama

Dr. Stewart D. Redwood

Cecropia Press

Photographs:
Jose Oller Collection, Panama, courtesy of Germinal Sarasqueta.
Photos taken by Leo George Inglis.
Bert Harrison Collection, Australia courtesy of Peter Tristram.
Vicente Pascual Collection, Panama.
Photos published in several contemporary articles about the mine (Chibas, 1898; Woakes, 1899, 1923; Goldman, 1912).
Kevin Heather Collection (1987), La Serena, Chile.
Carl Nelson Collection (1991), Boulder, Colorado.

Front cover photograph: Bert Harrison Collection.
Back cover photograph: Jose Oller Collection.

972.87
R249 Redwood, Stewart D.
Pirates of the Darien : the history of the Espiritu Santo de Cana Gold Mine / Stewart D. Redwood ; prologuista Germinal Sarasqueta Oller. -- Panamá : Cecropia Press, 2026.
240 páginas ; 25 cm.

ISBN 978-9962-715-66-5 (Tapa suave)

1. PANAMÁ - HISTORIA I. Título.
2. Tropical America 3. Tropical History 4. Central America
5. Mining 6. Precious Metals 7. Gold 8. Natural History

Cecropia Press
cecropiapress.com

This book is dedicated
to my dear wife Maite,
my treasure from Bolivia.

Pirates of the Darien
The History of the Espiritu Santo de Cana Gold Mine, Darien, Panama

Contents

Prologue 15
Introduction 21
1. Location and Topography 25
Location 25
Topography 26
Hydrology 26
Climate and Vegetation 27
Population and Political Jurisdiction 27
2. Access 29
The Historical Transport Route 29
Schooners 29
Steamers 31
Motor Vessels 33
Pack-animals and Wagons 33
Narrow Gauge Railway (1907) 36
Bullion Shipment 37
Airstrip 39
Access Today 39
Santa Cruz de Cana Town 40
Travelers' Tales 41
Eduardo Chibas (1898) 41
Robert Williams (1908) 43
Edward Goldman (1912) 44
Seth E. Meek and Samuel F. Hildebrand (1912) 45
3. Geology 47
4. Prehistory 51
5. Spanish Gold Mining 53
Early Spanish Gold Mining in Darien (1513-1559) and Veraguas (1524-1620) 53
The Discovery of Santa Maria (1665) and Cana (1689) 53
Captain Bartholomew Sharp's Raid on Santa Maria (1680) 54
Santa Maria Seized by Peter Harris (1684) 59
William Dampier's Description of Santa Maria (1694) 60

William Paterson's Description of Cana (1700): Scots Colonial Dreams 63
Nathaniel Davis's Raid (1702) 67
Robert Allen, a Scots Deserter Cum Pirate (1702) 68
Charles Tibón's Raids (1712, 1724) 69
Mine Collapse, Cuna (Guna) Revolt and Mine Closure (1727) 69
Darien Province (1774) 70
Don Andrés de Ariza's Reports (1774, 1778, 1780) 71
Spanish Mine Description 76
Historical Maps 77
6. Explorers in the 19th Century 79
Emile Lebreton (1840-1852) 79
Irish and English Explorers: Edward Cullen (1852) and Lionel Gisborne (1853) 79
Dr Wozencraft (1870s) 80
Lucien Napoleon Bonaparte Wyse (1876-1877) 80
Armand Reclus (1876-1877) 83
The Gold Mines of Cana Company (1878-1880) 85
7. The Lost Gold Mine Reopened 85
Compañía Minera del Darien, (1883) 85
Don Francisco B. Rojas (1870-1883) 86
Rediscovery (1883) 86
The Darien Gold Mining Company Limited (1887-1891) 88
Vicente Restrepo Maya (1837-1899) 90
Saturnine Maisounabe 92
Richard Hammersley Heenan (1847-1920) 93
Ernest Woakes, FIMM (1864-1947) 93
The Darien Gold Mining Company Limited (1891-1907) 95
The Rediscovery of Espiritu Santo Mine (1893-1896) 96
José Oller Navarro (1882-1984) 102
Bert Harrison 103
Darien Monetary Tokens 103
The Thousand Day War (1899-1902) 105
The Independence of Panama (1903) 105
William E. Proctor (1903-1907) 109
Block Caving (1904) 110
A Mine Tour 111

8. The French Mine (1905-1912) *115*
French Coup (1905) *115*
The Darien Gold Mining Company Limited (1907-1912) *116*
Placer Gold Mining and Disseminated Gold *120*
Rubber Plantations *121*
R. Williams, Botanist (1908) *123*
E. Goldman, Zoologist (1912) *123*
9. Gold Production *127*
Spanish Period (1675-1727) *127*
English Period (1887-1912) *128*
Summary *130*
10. Arias Ownership (1915-1927) *131*
Don Pablo Pinel (1869-1951), Legal Representative and Liquidator *131*
Don Tomás Arias (1856-1932), Founding Father of Panama *131*
11. The Panama Corporation Limited (1926-1948) *133*
Company History *133*
Move to Canada (1932) *134*
War Casualty: Gold Mines Closure (1942) *135*
Suspicious Americans *135*
Cana Exploration (1927-1932) *137*
Stanley Low, MIMM *138*
Duncan Elliot Alves *141*
Hugh Marriott, MIMM *144*
Walter Broadbridge, MIMM *144*
12. Exploration (1958-1996) *147*
Louis Martinz (1950s) *147*
South American Gold & Platinum Company (1958) *148*
Mauricio Hochschild Mining Group (1963-1966) *148*
Texas Gulf Inc.(1970s) *149*
IXTAPA (1979-1993) *149*
Freeport Exploration Company (1985-1988) *150*
Gold Fields Limited of South Africa Ltd (1992) *150*
ANCON (1993-1996) *150*
Link Mineral Ventures Limited (1995-1997) *151*
Darien National Park *151*

José Oller Collection 153
Vicente Pascual Collection 185
Kevin Heather Collection 191
Bert Harrison Collection 199
Historical Collection 209
Afterword 217
Acknowledgments 219
Sources 221
References 223
End Notes 231
Author Biography 237

Espiritu Santo de Cana Gold Mine

View of the Cana mine camp and the Cana Plateau to the east.
José Oller collection.

Prologue

In this province of Darien there are many very rich mines discovered, there is gold in great quantity, twenty rivers have been discovered and thirty that have gold come out of a mountain range that is up to two leagues from this Villa the Real of Santa Maria.

—Vasco Nuñez de Balboa, letter to the King Ferdinand II, 1513

Finally, the history of the legendary Espiritu Santo gold mine and its nearby deposits located in the Pirre mountain range and the so-called Cana plateau, in the province of Darién, has been written.

I am filled with joy and very grateful that my learned friend of many years, Stewart D. Redwood, a Scotsman resident of Panama City since 1994, has offered me the opportunity to preface this essay. Dr. Redwood holds a doctorate (PhD) in geology and has more than four decades of working around the world in mining and discovering metal deposits. His specialty is gold and copper in Latin America and the Caribbean. He has also written numerous geological reports and papers, as well as several archaeological papers with the Smithsonian Tropical Research Institute in Panama (STRI). His magnificent work is entitled Pirates of the Darien: The History of the Espiritu Santo de Cana Gold Mine.

Without a doubt, Stewart Redwood is the ideal person to leave us this unique essay on the human and industrial history of the Colombian era and our new republican era on the distant plateau of Cana in the Pirre mountain range, Darien province.

I have oriented this preface in four periods.

- The great legendary period of the seventeenth and eighteenth centuries: the Spanish discovery of the Holy Spirit of Cana and the brutal reality of mining in the green "hell" of the Darien.

- The renaissance of the late nineteenth century: the arrival of the English (The Darien Gold Mining Co. Ltd.) and the industrialization of the jungle.
- The first family tie (1903): the time of my grandparents, great-grandfather and great-great-uncle, the transition from Colombian to Panamanian sovereignty and the harshness of mining at the beginning of the twentieth century.
- My personal experience (1975-1980): reaching the ruins to bring life back to the mountain.

My enthusiasm and fondness for the Cana mines was born from a very young age when my parents told me about the participation of my ancestors in The Darién Gold Mining Co. Ltd. Ever since those days I dreamed of reaching those majestic mountains dense with tropical forests, almost unexplored and abundant in mysterious legends.

The British, pioneers of the industrial age, arrived on the isthmus in the late nineteenth century attracted by the mining wealth of Cana. It was through my maternal lineage that this story came to life: my great-grandfather, Prospero Pinel, along with his younger brother Pablo and partners from Manchester—who were already marketing the pearls of the Pinel Brothers' pearl-diving business—reorganized the concession for the mine in 1902.

Pablo Pinel was the legal representative. My grandfather José Oller Navarro (1882-1984), "Papiche," as his grandchildren called him, worked from a very young age as a messenger and assistant accounting clerk at Pinel Brothers, and he was sent to the mine to look after the interests of the company. Grandfather Papiche, who was not yet Prospero Pinel's son-in-law, spent November 3, 1903, in Cana. During his stay in that mountainous jungle he overcame challenges, adventures and even managed to descend the almost three hundred meters depth of the Espiritu Santo mine. In 1925, he narrated for posterity the descent of the 300 meters in his essay "The Mining Industry in Panama," a work based mainly on the deposits of the Cana plateau, with a prologue by his friend Guillermo Andreve (1879-1940). Fifty years later he published the second edition.

An endearing anecdote from those years occurred while my grandfather, José Oller, was writing his work on mining in Panama in 1925. My mother, Rosario Oller Pinel, who was only nine years old at the time, asked him curiously, "Dad, where did you learn to speak English?" He answered simply, "In the Darien, my little daughter." Astonished, she replied: "Over there? But only indigenous people live there." Papiche ended up explaining to her laughingly that he had learned it living with the English engineers and administrators in the famous mines of Cana.

On the other side of the family, my paternal grandfather, Pedro Sarasqueta Ugarte (1880 -1945) a native of Portugalete, Vizcaya, Spain, and a builder by profession, won a contract in 1907 to build houses for the operators of the new gold deposits within the valley of the Cana River. Thus, after a journey with many pitfalls and hardships, made together with his small family of his wife Dominga Castillero and little daughter Libertaria, he arrived at the town of the famous gold mines.

However, in that beautiful fertile valley, there was a dangerous environment, with assaults and harassment and insufficient authority. Already working and protecting his family, the miracle of San Antonio happened.

In the middle of the Darien jungle, yellow fever viciously attacked those who ventured through it. Foreigners were more vulnerable. It was said that 50% of those who got sick did not survive. Don Pedro was attacked by the dreaded fever. His wife felt helpless and vulnerable with their little daughter and, while she was busy caring for the sick man with the precarious remedies she possessed, she remembered an old revolver that Don Pedro possessed. In those moments of anxiety and pessimism, the revolver restored some tranquility. The love for her spouse, her great faith in God, and her devotion to St. Anthony, who heard her prayers, saved her husband's life. They then returned to the city as soon as possible.

I first visited the Darien in 1961 when I was 12 years old. I flew in a G73 Gruman seaplane, registration HP-383, with my mother who was the lawyer for the newly-opened Panama Fishing Club later called the Tropic Star Lodge, in Piñas Bay, Jaqué, Darien province. From the plane I saw the imposing Pirre mountain range for the first time, and although I couldn't make out the Cana Plateau, I felt very close to and attracted by the legendary place.

In 1975, I finally managed to visit Cana to investigate a plane crash at the grass landing field, which has a steep slope and is surrounded by high hills. I felt happy after fulfilling my responsibilities, and hurriedly toured the area, and appreciated all the abandoned industrial archaeology and some nearby hamlets. I even managed to rescue several antiquities such as crucibles used to amalgamate mercury with gold to clean it; and old bottles of the famous green balsamic liquid called Bay Rum, used in abundance to soothe the itching of mosquito bites, gnats or *chitras*, which in Darién Emberá are known as *morrongoi*. I also collected several horseshoes used by the mules in the so-called blood train. My grandfather Papiche told me more than half a century ago that those mules came from the Chilean port of Coquimbo north of Santiago.

Five years later, at the beginning of 1980, I was thrilled to visit Cana several times by helicopter and in a Britten Norman Islander, the HP-870, with my friend Gabriel "Bibi" Lewis, whose company IXTAPA S.A. had obtained the concession of the Cana mines. Its directors were Gabriel Lewis Galindo and the Spaniard Manuel Hernández de los Ángeles. To the latter I tried to sell an old Jeep CJ6 with a long wheelbase which had recently been rebuilt with a winch. Don Manuel told me: if you get it to the Cana mine, I'l buy it from you and I'll also pay for the transport. In fact, I immediately realized that the only way to get that Jeep there was using the Short Brothers SC7 "Skyvan" with a turbo fan, registration FAP-300, with a large rear door and platform, recently acquired by the former Panamanian Air Force. Through Captain Rodolfo Castrellón, Bibi's brother-in-law, and our friend Juan Méndez Salazar, I managed to send the jeep

for free and of course I did not charge for the transport. It remains there like the mine: abandoned, corroded and stripped of many of its parts. I keep a photo of its current appearance as a reminder.

Many other Panamanians from the capital went to work in Cana. On a list that appears in my grandfather's essay are: Alejandro Ayala, Pedro Campagnani, Joseph Cánepa, Nicolás Ossa, José Dolores Ossa, Emeterio Ossa, Juan Manelia, Atilio Repetto and many others that José Oller did not remember.

To perpetuate the legacy of these Panamanians who braved the anguish and isolation of the Cana mines, an association of their descendants was recently created. It is a tribute to those who, in search of a dignified living, gave their maximum effort in one of the most remote and dangerous places in the country.

Thank you Dr. Redwood for your excellent essay. In Darién there is a saying:

When you go to the Darien, entrust yourself to Mary.
With her is your entry ticket, and with God your way out.

Germinal Sarasqueta Oller

Introduction

This book tells the fascinating and dramatic story of the biggest gold mine in the Spanish Empire and probably, at the time, the world. The famous Espiritu Santo de Cana mine – the Holy Spirit of Cana – lies deep in the Darien jungle of Panama. The story involves Spanish conquistadors, African slaves, Scottish colonizers, Cuna (Guna) Indians, English and French pirates, an indigenous uprising, rediscovery by a Colombian historian, reopening of the mine by a company from Manchester, England, West Indian miners, French capitalists, flooding, civil war, secession, and a founding father of Panama. In the 20th century the mine was explored by the Panama Corporation of England, whose story is also told, and later by several international mining companies.

The mine is situated in a remote part of the Darien Gap close to the present-day Colombian border in the rain forest of the mountains of the Pirre Massif. The mine district was not discovered until relatively late in the Spanish colonial period in about 1675 and it was mined for about 52 years until 1727 when it was abandoned after a tragic cave-in, constant pirate raids, and an Indian uprising.

It lay hidden in the jungle again for more than a century and a half. Lucien Napoleon Bonaparte Wyse, great nephew of Napoleon, tried to stake it for the French Canal Company but it was a Colombian historian, Vicente Restrepo, who researched the mine's "romantic" history and got title to the mine. He brought in an English company, the Darien Gold Mining Company Ltd., with Anglo-French capital, to mine it in 1887. It rediscovered the overgrown, collapsed and flooded Spanish mines and mined gold again for 25 years from 1887 to 1912.

The mine was bought by Don Tomas Arias, one of the founding fathers of Panama. In the 20th century the Cana mine was explored by the Panama Corporation of England and Canada, the most important gold mining company in Panama between the World Wars. After World War Two, the Cana mine was explored by modern techniques for gold and copper by several major mining companies from around the world and, while new resources were found, they did not warrant building a new mine. This important historical and industrial archaeological site is now overgrown and forgotten in the Darien National Park.

The history of the mine is summarized in Table 1.

Year	Owners and Events
1665	El Real de Santa Maria was founded. Placer gold mined on River Tucuti.
~1675	Espiritu Santo de Cana discovered and mined by Spaniards (exact date not known).
1680	Captain Bartholomew Sharpe, an English buccaneer, raided Santa Maria for gold that was brought from Cana, which was already in full operation (book description).
1684	Captain Peter Harris, an English buccaneer, raided the El Real mine (book).
1700	William Paterson's Prospectus for the Scots Darien Company described the Cana gold mine (first description of the mine).
1702	Nathaniel Davis, an English buccaneer, raided the Espiritu Santo mine (book).
1712	French pirate raid of Santa Cruz de Cana by Charles Tibou.
1724	French pirate raid by Charles Tibou.
1727	Espiritu Santo de Cana mine closed due to collapse of the shaft with two deaths.
1727	Cuna rebellion led by mestizo Luis Garcia forced the Spaniards out of Darien.
1727	The last owners of the mine were Don Antonio Arguelles, Don Diego de la Guardia, Don Juan de Leon, Don Diego Mogica and Don Antonio de Sosa.
1727	Don Juan de Ibiricu reopened the mine but did not have title and had to stop.
1727-1883	No mining activity.

1727-1883	No mining activity.
1774	Report on Cana mine by Don Andres de Ariza, Governor of Darien.
1778	Rebellion quelled and Darien pacified, by which time it was depopulated.
1779-1839	No activity.
1840-1852	Emile Lebreton formed a French company to reopen the mine.
1870s	Dr. Wozencraft, a spiritualist, tried to find the mine.
1877	Lt. Lucien Napoleon Bonaparte Wyse visited the mine and claimed it for the French Canal Company.
1878-1880	The Gold Mines of Cana, New York tried to reopen the mine.
1882	Staked by Compañía Minera del Darien, Bogota (Vicente Restrepo).
1882	30 December, contract of purchase by S. Maisounabe, Paris of 4 lode claims and 3 placer claims.
1883	Don Francisco B. Rojas rediscovered the mine and took legal possession for the Compañía Minera del Darien. Expedition 6 July to 10 December 1883.
1884	23 January. The 4 lode claims were granted.
1884-1886	3 year gap in Cia Minera del Darien history presumably while S. Maisounabe was trying to raise finance in France.
1887	20 February. Agreement with Saturnin Maisounabe to buy the mine. G. M. Braggioth and A. Figge to form a new company in London, Maisounabe to get 40% of the new company. 25 March, Heenan added and got 2/8th share of the new company.
1887	2 March, Darien Gold Mining Company Ltd., Manchester formed. 3 March, deal with Braggioth, Figge and Heenan to buy the mine
1887-1892	Mined the El Rey Mine (North Mine). Installed a 20-stamp mill.
1891	The Darien Gold Mining Company declared voluntary liquidation. A new company was formed with the same name.
1893-1907	Espiritu Santo Mine (South Mine) rediscovered in 1893 and mined profitably from 1896. Installed a 40-stamp mill.
1899-1902	The Thousand Day War, a Colombian civil war between ruling Conservatives and rebel Liberals, devastated Panama.
1903	The mine ceased to be in Colombia and became part of the new Republic of Panama.
1907	The mine flooded and production dropped.
1907	The Darien Gold Mining Company declared voluntary liquidation.
1907	New company formed, The Darien Gold Mining Company Ltd., with French capital, directors and management.

1910	Naturalist Robert Williams used the mine camp as a field camp and described the journey.
1912	Biologist Edward Goldman used the mine as a field camp during the Smithsonian's Biological Survey of the Canal Zone and described the journey.
1912	Mining stopped.
1913	Company in liquidation.
1915	Assets bought by Tomas Arias. The mines were excluded. Arias carried out ranching.
1927-1932	Explored by the Panama Corporation, Limited of England with a 3 year option. Evaluated bulk mineable target on Cerro San Jose.
1932-1964	No activity.
1950s	Mine concesion acquired by Louis Martinz.
1958	South American Gold & Platinum Company evaluated the placer gold deposits.
1964-1966	Mintrade Group (Mauricio Hochschild company) exploration.
1970s	TexasGulf Western Inc. exploration. Drilled 16 holes.
1970s	St. Joe Minerals exploration.
1980	Darien National Park created. Formerly Alto Darien Protection Forest 1972. Became a UNESCO biosphere reserve in 1983.
1980	Sociedad de Inversiones IXTAPA S.A. was awarded the concession.
1985-1988	Optioned to Freeport Exploration Company. Drilled 46,000 m in 40 holes at North Mine.
1992	Evaluated by Gold Fields Ltd of South Africa. 1,200-hole soil auguring program.
1993	IXTAPA bought by NGO ANCON, Panama.
1995-1997	Link Mineral Ventures Ltd., Toronto explored the project under an option from ANCON.
1996	Mining was banned in the Darien National Park.

Table 1. Summary of the history of the Espiritu Santo de Cana gold mine.

1. Location and Topography

Location

The Cana mine is located in the Pirre mountains (Serrania de Pirre) 240 km southeast of Panama City and only 5 km northeast of the Colombian border at Alturas de Nique. The coordinates are 7° 45' 31" N, 77° 40' 50" W (WGS84, UTM Zone 18, 204,300 east, 858,500 north).

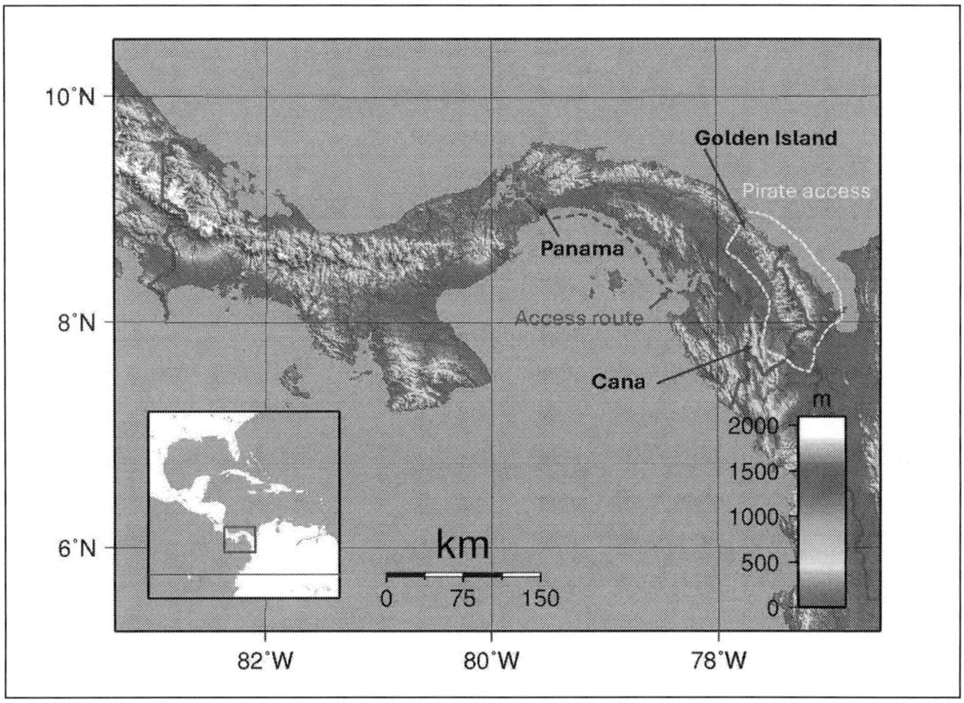

Location map of the Espiritu Santo de Cana mine.

Topography

Cana is located on the eastern side of the Pirre mountains (up to 1,654 m altitude) on the edge of a wide, flat intermontane basin called the Cana Plateau, that is about 12 km long in a north-south direction by up to 7 km wide with an altitude of 470 m. The plateau is bounded on the east side by the Setutele mountains (1,074 m) and to the south by Alturas de Nique (1,674 m). The mountain ridges mostly trend north-south and there is high relief of about 1,200 m.

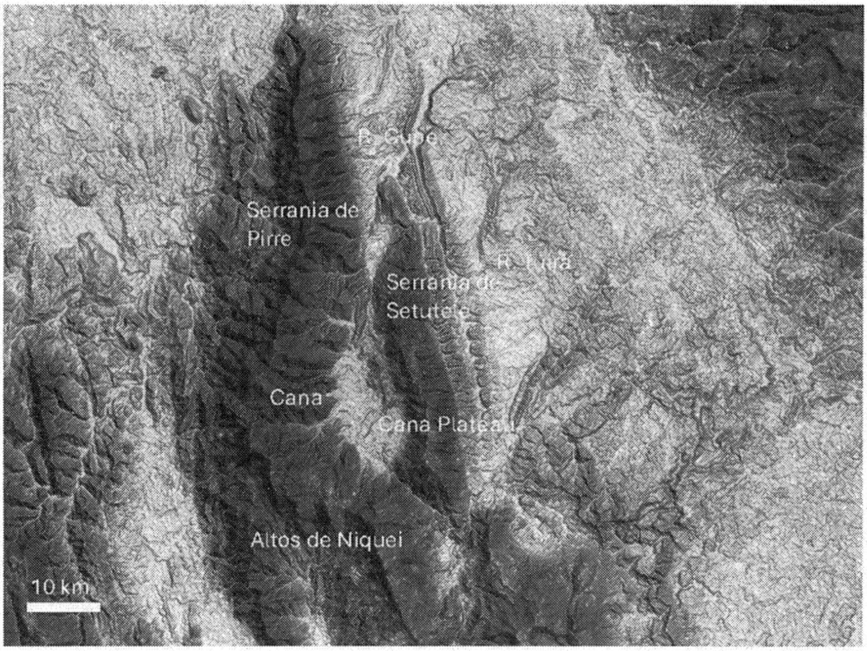

Digital elevation model (DEM) of the Cana district based on satellite radar mapping to see through the cloud and forest canopy ((SRTM, 12.5 m resolution).

Hydrology

The Cana valley is drained to the south by the headwaters of the River Tuira. The River Cana is a very short (~9 km), easterly flowing river that drains the Serrania de Pirre and then passes through the mine into the Cana plateau to join the River Seteganti from the SW, also very short at ~9 km, which then joins the Tuira, also called the River Grande (Mendez, 2004, p. 57). This crosses the Serrania de Setetule range on the south side of the Cana valley by following a prominent NW-SE-trending fault that forms a deep canyon with the Peña Hueca waterfalls that drop 200 m. It is joined by the River Mangle on the right bank and then flows north as the River Tuira gathering size, then northwest as a major river into the Gulf of San Miguel.

The Serrania de Setutele is drained by the River Paca which flows NE to join the River Tuira. The transport route to Cana by mule trail and light railway was by the NE-trending valley of the River Cupé further north to Boca de Cupé where it joins the Tuira.

Climate and Vegetation

The vegetation is pre-montane wet forest with annual precipitation greater than 2,200 mm (Boca de Cupé weather station; Bush & Colinvaux, 1994) to 3,000 to 3,500 mm (Piperno, 1994). There is a dry season from January to April with only 200 mm of rain.

Population and Political Jurisdiction

Santa Cruz de Cana is situated in the Corregimiento of Boca de Cupé, District of Pinogana, and Province of Darien. There are no inhabitants in Santa Cruz de Cana nowadays. The nearest settlements are at Boca de Cupé, 33 km north and at El Real de Santa Maria, 42 km to the north. The population of the corregimiento of Boca de Cupé is 1,117 (INEC census, 2023)

2. Access

The Historical Transport Route

Access to the Cana mine was always difficult and involved maritime, fluvial and mule transport, and later a light railway. From Panama City transport was by boat across the Gulf of Panama and Gulf of San Miguel, then up the River Tuira to El Real at the confluence with the River Chuqunaque, near the tidal limit. This took 5 to 12 days by coastal schooner, and 24 hours by steamer. From El Real people and cattle took a mule trail to Cituro on the River Cupé, and then a road to Cana which took 2 days. Goods and machinery were taken from El Real further up the River Tuira to the mouth of the River Cupé at Boca de Cupé, and then up the River Cupé to Cituro in long dug-out canoes called piraguas or bogas. Transport from Cituro to Cana was by pack-mule, pack-ox and wagons. A narrow gauge steam railway was built from Boca de Cupé to Cana in 1907.

Schooners

Schooners took 5 to 12 days to make the voyage from Panama to El Real. A normal voyage took 2 days to the mouth of the River Tuira, then sailed up-river at 6-hour intervals with the incoming tide, and anchored for the rest of the time, to reach El Real on the fifth day (Chibas, 1898).

Woakes (1923) described the long journey to the mine by schooner.

> In the early days Cana was very difficult of access. The only route was by means of small coasting sailing vessels from Panama. It was a tedious voyage of six or seven days, cooped up on the deck of a dirty little schooner in the broiling tropical sun, often becalmed for a whole day.

From	To	Route	Method	Distance - time
People				
Panama	El Real	Gulf of Panama, River Tuira	Schooner or steamer	5-7 days by sail. 24 hours by steamer.
El Real	Cana	Trail via Cituro, Paca	Mule	2-3 days
Railway, 1907				
El Real	Boca de Cupé		Mule	
Boca de Cupe	Cana	River Paca valley	Railway	47.5 km
Goods, machinery				
El Real	Boca de Cupé, Cituro	River Tuira, River Cupé.	Piragua (goods, machinery)	
Cituro	Cana	Trail	Pack-mule, oxen, wagons (goods)	

Table 2. Summary of historical access from Panama to Cana.

Name(s)	Type	Built	Dimensions	Capacity	Disposed of
Darien *Panama* (1901) *Darien* (1902)	s.s. (single screw steam ship)	England, 1897. Completed Panama, 1898.	60' x 10'	20 tons	1901 Colombian Navy gunboat ren *Panama*. 1901 was captured by rebels *Darien*. 1902 Returned to Colombian Navy. 1903-04 Panamanian Navy. Fate unknown
Cana	s.s.	Probably England 1901/02.	56.4 x 12.15 x 5.0 ft, steel		Acquired by Arias with mine in 1915. Operated until 1930? Abandoned in mouth of River Platanal, remains still visible in the 1970s.
Tuira (Tuyra)	s.s.	Probably England 1901/02.	Steel	7 tons	Commandeered by US Marines in 1903-04. Listed in 1915 sale inventory. Operated until 1930? Abandoned at Playa Trujillo, Panama City, remains visible until the 1970s.

Table 3. Steamers of the Darien Gold Mining Company Ltd.

> Once the mouth of the Tuvra River was reached, in the Gulf of San Miguel, the sails were taken in and the sweeps got out. About six tides were sufficient to take the boat up to the village of El Real, where tidal water ceased. Between the tides the boats anchor and the passengers are at the mercy of the mosquitoes and sandflies.

The type of coastal vessels were *balandras* (sloops, 1 mast), *lanchas*, and larger *goletas* (schooners, 2 or more masts).

Steamers

A small screw steamer called the *Darien* was built for the Darien Gold Mining Company Ltd in England in 1897 and started operating in March 1898. She was built to the maximum dimensions that could be carried over the Isthmus on the Panama Railroad. Her length was about 60 feet and breadth 10 feet with capacity for 20 tons of cargo. The engines and boiler were installed after launching in Panama. She operated from Panama City across the Gulf of San Miguel and up the Tuira River to Real de Santa Maria, which took 24 hours. She was taken over by the rebels during Thousand Day War (1900-1902) and was used in the rebel naval fleet at Panama (Chibas, 1898; Woakes, 1923). A Navy list gives the name as *Panama*, ex-*Darien*, (she appears to have been renamed but reverted to *Darien*) that was a gunboat bought by the government on 17 June 1901, seized by the rebels on 4 November 1901 *(Darien)*, returned to the government after the war on 21 November 1902, and was transferred (*Darien*) to the Panamanian Navy on 6 November 1903.[38] The Panamanian army and navy were abolished by law on 20 November 1904 and the navy ships were sold, but the fate of the *Darien* is not known.[39]

She was replaced two other small steamers called *Cana* and *Tuira* (or *Tuyra*). The *Tuira* was commandeered by the US Marines in 1903-04 in the aftermath of the independence of Panama. Both steamers are listed in the sale inventory of 1915, which is the last reference to them (Mendez, 2004). It is not known when they stopped operating. Raúl Arias de Para, great grandson of Tomas Arias, has

Above: The steamer *Darien* in 1898 (Chibas, 1898).

Below: The steamer *Cana* in February 1912 (Goldman, 1912). [86]

two wall clocks from the company's steamer *Cana*; one is in his house in El Valle, and one is in the library of the Canopy Camp Darien eco-lodge, near Meteti.[3] The *Cana* was abandoned at the mouth of the River Platanal in Garachine Bay, Gulf of San Miguel, and the remains were still visible in the 1970s, while the *Tuira* was abandoned at Playa Trujillo (now the site of the Fish Market and Fiscal Pier) in Panama City and the remains eventually rusted away in the 1970s (Mendez, 1979, 2004).

Motor Vessels

Sheridan (1926) wrote:

> All of the quartz and placer gold properties can be reached from Panama City by gasoline or oil-burning boats, to the Gulf of San Miguel, and then up the Tuyra River to Real de Santa Maria and Yaviza. The U. S. Gunboat *Yorktown* made the trip as outlined above, and anchored off Yaviza for some time, landing 250 marines, during a rebellion against Colombia in 1903.

Pack-animals and Wagons

From El Real people and cattle took a 20 to 25 mile mule trail to a half-way station at Cituro on the River Cupé (Woakes, 1899, 1923). From Cituro to Cana there was a road of 19 miles with steep grades and more than a dozen river crossings until the Cana Plateau was reached. There was a half-way house at Paca station. The worst sections were graded and "macadamized" (asphalted) in the late 1890s (Woakes, 1899, 1923).

Chibas (1898) described the route in detail. He was contracted to survey a railway or paved road.

A 50 km narrow gauge steam railway was built from Boca de Cupé to Cana in 1908.

Goods and machinery were taken from El Real further up the River Tuira to the mouth of the River Cupé (Boca de Cupé), and up the Cupé to the limit at Cituro station, using long dug-out canoes called piraguas or bogas. The trip took three to eight days, and the return in twelve hours with a good river (Woakes, 1899). Transport from Cituro to Cana was by pack-mule and pack-ox which could carry loads of 130 pounds and 200 pounds respectively. Wagons hauled by oxen were used in the dry season from the late 1890s, but the road was a swamp during the nine-month rainy season. All heavy transport for the year had to be done during the three month dry season. All machinery had to be broken-down or sectionalized for transport (Woakes, 1923).

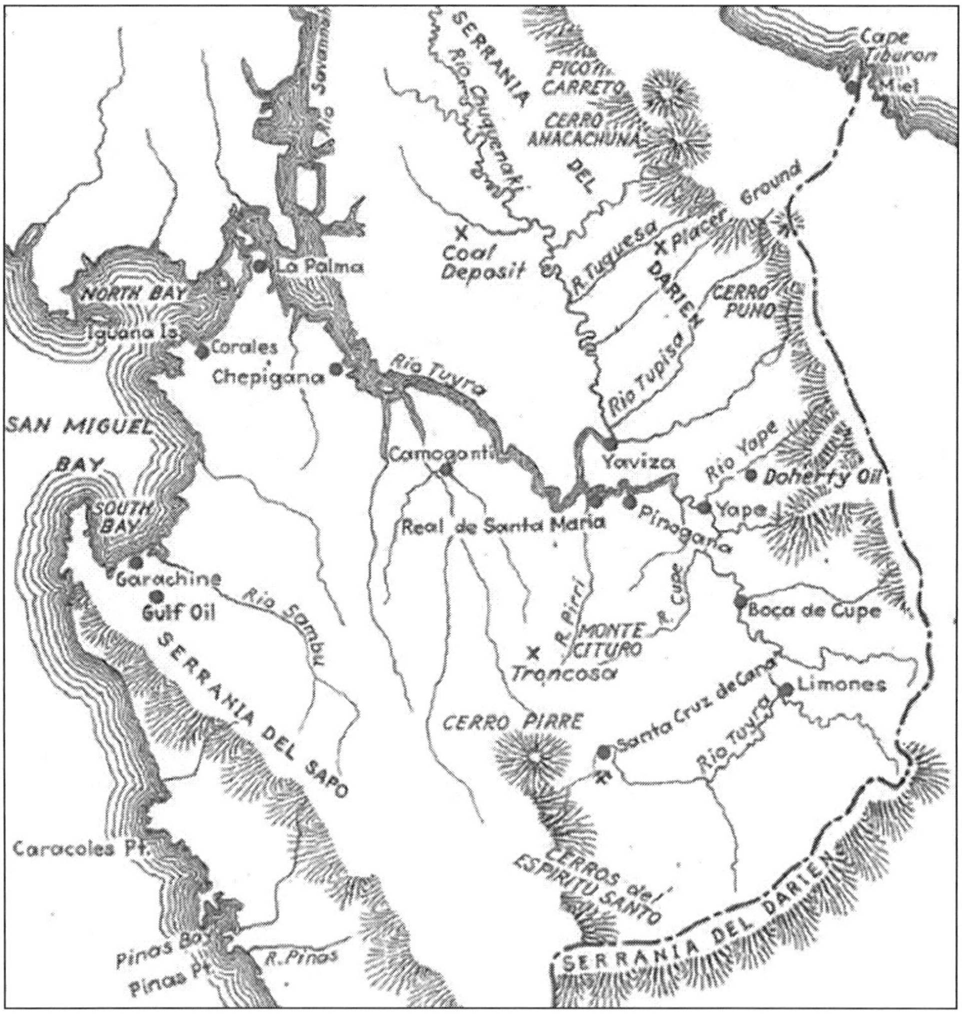

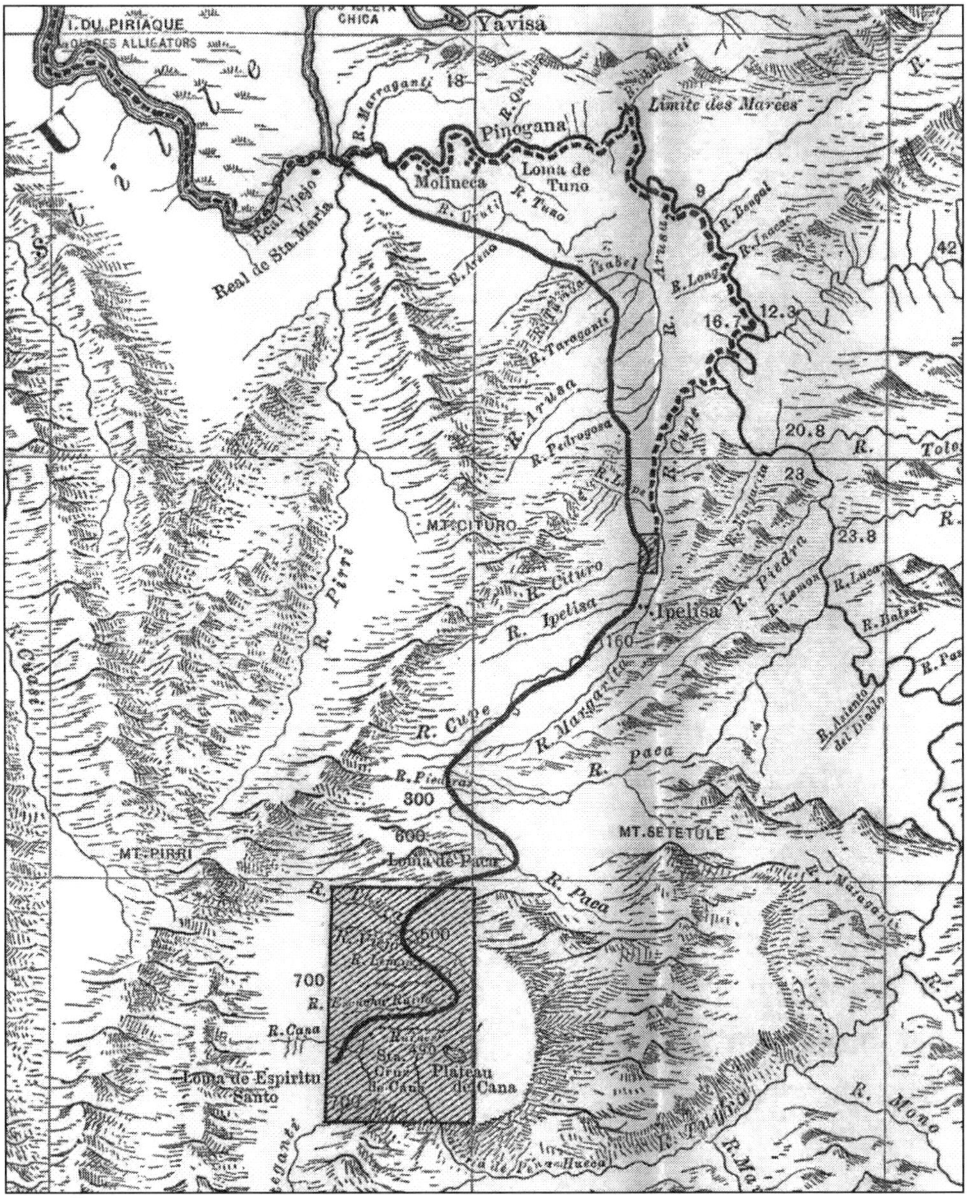

Above: The transport routes from El Real to Cana (Woakes, 1899, detail).

Opposite page: Darien map (Sheridan, 1926).

Narrow gauge Railway (1907)

The first act of the French directors of the renovated Darien Gold Mining Company of 1907 was to build a 47.50 km long, 0.60 m narrow gauge railway from Boca de Cupé at the limit of navigation on the Tuira River to the mine at a cost of 3 million Francs, the equipment having already been bought by the previous company. The two steam engines were manufactured by Orenstein & Koppel at Drewitz in Potsdam, near Berlin, Germany, a specialist in light railway locomotives. The steam locomotives and wagons can still be seen at the site. From Boca de Cupé the railway passed through the settlements of Bajo Grande, Buena Vista, Quebrachonta (Quebrada Chonta), Cituro, Aguas Buenas, Paca, Escucharruido and San José, all set up by the company (Mendez, 2004). The railway was described as precarious and the bridges as dangerous, and derailments were common with injuries, mutilations and deaths (Mendez, 2004, p. 56-57).

Opposite page: The industrial archaeological heritage of the Cana mine: steam locomotives and pumping engine. A, B) Two abandoned light steam locomotives from the narrow gauge railway from Boca de Cupé to the mine built in 1907, manufactured by Orenstein & Koppel of Germany at Drewitz in Potsdam, near Berlin, a specialist in light railway locomotives. The larger engine (A) is 30 to 40 hp, and the smaller one (B) is 20 hp. C) A steam-powered pumping engine manufactured by Fraser & Chalmers, Limited at Erith, Kent, England. There are also winch drums from the shaft and railway carriages at the site. D) Advertisement by Fraser & Chalmers for a similar steam engine. Photographs taken in 2009 by Dr. Glyn Dawson, Chicago and published with his permission. Thanks to Jeremy Martin, London for identifying the locomotives. Plate from Redwood (2020). Engine (A) has a name plate Cituro according to Arauz Torres (2014: 352) who reproduces a photo of it taken by his father in 1962 with Merardo Murillo, the last surviving miner, and another photo taken with him in 2008.

Bullion Shipment

The gold bullion was shipped under strict security by the Darien Gold Mining Company Ltd from the Administration Office at Cana by the steamer *Cana* by the River Tuira and the Gulf of Panama to the company's legal representative in Panama City, Don Pablo Pinel. He delivered it to the Agent of the Royal Mail Steamship Company with a Bill of Lading, to be sent by railway to the port of Cristobal, then by steamship across the Atlantic Ocean to England and delivery to the company in Manchester or the London exchange (Oller, 1975).

A photograph of the gold production for the month of November 1904 shows 27 gold ingots with a total weight of 6,000 Troy ounces, then valued at £154,418 and today worth about £21.6 million or US$29.4 million (at $4,900 per ounce).

Above: Aerial view of the Cana airstrip and the camp in 1991 (Carl Nelson).

Below: Unloading supplies from a Cessna at the Cana airstrip in 1991 (Carl Nelson).

Airstrip

Access to Cana in more recent times was by light aircraft to a grass airstrip built at an unknown date but probably in the late 1950s or early 1960s when modern exploration started. It is believed to be abandoned now. Carl Nelson visited Cana several times in the 1980s-early 90s by chartering a small plane in Panama City. He recalls that they carried with them a can of gas for the lawnmower at the camp in order to cut the grass on the airstrip before the plane could take off.

Access Today

Access to Cana today is even harder than it was in the past. There is no road access and the only options are by hiking for several days, or by helicopter which requires special permission as the border area is a No Fly Zone. The land access is by trekking for 2 to 3 days from Boca de Cupé, 33 km north or from El Real de Santa Maria, 42 km to the north.

Below: Departing from La Palma to explore the lower River Tuira by cayuco, 2015 (S. Redwood).

Exploring the lower River Tuira by cayuco, 2015 (S. Redwood).

Santa Cruz de Cana Town

Santa Cruz de Cana or just Cana was located in an enclosed plateau crossed by the River Cana, a tributary of the River Seteganti which joined the upper River Tuira, also known as the River Grande. Santa Cruz de Cana was a Spanish mining town from ~1675 to 1723. It is said to have had as many as 20,000 inhabitants. Later, it was an English and Antillean mining town from 1887 to 1913, and it was abandoned by 1924.

Nathaniel Davis, who raided the mine in 1702, wrote the first eye-witness description of the mine. He reported that the town of Cana, which they burned, had about 900 houses and but one church (Davis, 1704 in Joyce, 1934).

Cana was the first place to use hydroelectricity in Panama for lighting and to operate the gold mining plant. It had an aqueduct, hospital, commissary, school, Protestant church, municipality (Corregidor), jail, two villas of two-stories for the managers' lodging, dozens of houses, chales, bugalos and many camps with more than 60 rooms. There was a cemetery at San José on the north side of the

River Cana opposite the bridge and grazing for milk and beef cattle. Further north there was a sawmill at Escucha Ruido, powered by hydroelectricity, that supplied lumber for construction and supports for the mine. There was a rubber plantation with one million trees in the flood plain of the River Seteganti in the same plateau (Mendez, 2004).

The sale inventory of 1915 shows that there were 198 houses including 165 houses of wood with an iron roof, 31 houses of "caña y paja" (cane and straw), and 2 two-story houses, valued at £422 (Mendez, 2004, p. 59-62).[4]

The Antillean, Colombian and Indigenous miners stayed on after Arias bought the mine in 1915 in the hope that the mine would reopen and in the meantime grew tobacco, hunted and fished using dynamite. They gradually left and by 1924 the place was deserted. In 1926 a visitor reported that there were about 175 houses and much mining equipment, all idle (Sheridan, 1926).

Travelers' Tales

Eduardo Chibas (1898)

Eduardo Chibas was contracted by the Darien Gold Mining Company to report on the feasibility of building a railway or a 'macadamized road' to the mine in December 1896 (Chibas, 1898). He travelled from Panama to El Real by sloop in 5 days and then 2 days by mule to Cana. He described the route in detail:

> I arrived in Panama towards the latter part of December 1896, and at sunset one evening, shortly afterwards, I put out for Darien. At dawn of the next morning our sloop was sailing on the calm waters of the Pacific and passing close to the Pearl islands, so called on account of the large number of pearls covering the surrounding sea-bottom. The shells are secured by divers, who make a specialty of the work. Although the majority do not contain pearls, they are all valuable as mother-of-pearl. At the end of the second day we were entering the extensive gulf of San Miguel where the Tuira river empties into the

Pacific and about one hundred miles south-east of Panama. While sailing on the majestic Tuira, at the mouth of which there are no bars to obstruct navigation, we were able to make headway only at six-hour intervals, that is, while the tide was running in. The rest of the time it was necessary to anchor, to prevent the sloop from drifting back with the strong current caused by the outgoing tide. The swiftness of the current in either direction-due to the tide, whose effects are felt in a diminishing ratio for sixty· miles up the river-will be easily understood when it is known that the difference between high and low· tide in the gulf of San Miguel is twenty-four feet. Being forced to these frequent stops, two of which were made in front of native villages, our progress was naturally slow. We passed swarms of alligators, some of them from fifteen to twenty feet long. On the fifth day from Panama we reached Real de Santa Maria, the end of navigation, more than fifty miles from the mouth of the river. Near Real de Santa Maria there can be seen the ruin of the old town built by the Spaniards in the sixteenth century, with the forts they used to defend it from the repeated attacks of the Indians and pirates.

On the morning after our arrival we started on the inland journey, and, after nearly nine hours of hard muleback riding, over marshes and rugged hills and among colossal trees, we reached Cituro, the halfway station , where the night was spent. To this point the Cupé river, one of the affluents of the Tuira, is navigable by canoes during the rainy season, when most of the cargo is floated up. The loaded canoes sometimes make the trip in four days, but often require more than double that time. In the dry season scarcity of water in the Cupé river, makes navigation more difficult, and the canoes have to stop three miles below Cituro. In the rainy season the return trip from Cituro to Real is oftener made by canoe than overland. The thrilling experience of shooting down a few rapids below Cituro, and the changing tropical scenes at every turn of the winding river, add novelty· to this manner of making the return-journey, which takes from twelve to fifteen hours. The road from Cituro towards Cana crosses the Cupé river nine times, and runs up the Cupé hill, which rises abruptly about five

> hundred feet above its foot. Thence up and down a series of smaller hills to the crossings of the Paca river; thence up the steep Paca hill, towering nearly a thousand feet above its base. After reaching the summit, a rapid descent of four hundred and fifty feet took us to the upper Tuira, so close to its source that the river, at the crossing, is very shallow and hardly fifty feet wide. A ride through the Cana plateau, a picturesque and fertile valley a few miles in extent, and a further rise of two hundred and fifty feet along the Espiritu Santo range, brought us to the end of the road at the Cana mines. All the ascents and descents over the Cupé and Paca hills were made on very steep and long grades, varying from 20 to 27 feet in 100 feet! It is not surprising, therefore, that the cost of transportation over these twenty miles has been more than $50 in Colombian currency - equal to about $23 in gold - per ton. The cargoes are mostly carried on the backs of mules or oxen, which take at least days to make the journey during the rainy season. ·Wagons can be used only during two or three months in the dry season, when it requires the combined force of six mules and four oxen to haul one oi them, with a load of only fifteen hundred pounds, over the mountainous portion of the road, at the slow rate of ten miles a day. (Chibas, 1898).

He recommended the construction of a macadam road with a covering of broken stone or gravel 6 inches thick and 8 feet wide along a new route, suitable for wagons all year round, which was built in 1898.

Robert Williams (1908)

Botanist Robert Williams made the voyage to Cana in 1908 (Williams, 1908). The expedition sailed from Panama City on the steamer *Cana* at 6.30 pm on 1 April 1908 and arrived at the Gulf of San Miguel at 9.00 am the next day. They then sailed up the River Tuira near El Real. They accepted the invitation of the agent of the Darien Gold Mining Company, Mr. Lachszyrma, to stay at his station (or camp), called Marragantí, several miles upriver, where they remained for nine days making botanical collections. They then travelled to the mining camp at Cana, taking from 11 to 16 April 1908. The first two days travel was in a piragua carrying 1,200 pounds of iron rails to the company's first station above Marragantí, where they slept on a beach. On the third day they travelled to the third station, Cituro, on a narrow-gauge railway in the cabin of a 42 ton motor, having to walk the last two or three miles as the railway was not complete. On the fourth day they rode by mule to the third and final station called Paca, a distance of 13 miles. On the final day they walked the last remaining leg to Cana, a distance of 10 miles.

Edward Goldman (1912)

Zoologist Edward Goldman described the four-day journey from Panama to Cana in 1912 in order to carry out fieldwork for the Smithsonian Institution's "Biological Survey of the Canal Zone" (Goldman, 1912; Heckadon-Moreno, 2004b). He travelled from Panama City on the steamer *Cana* on the evening of 21 February 1912. By early morning they were at the mouth of San Miguel Bay. The day was spent steaming up the bay and the estuary of the River Tuira, with short stops at La Palma and Chepigana, "native villages, built mainly of palm-thatched houses picturesquely grouped on the southern shore." They had to anchor for an hour in the mouth of the River Chuqunaque at nightfall to wait for the tide and reached Marraganti at 9.30 pm where they were welcomed by the agent, Mr. Pedro Campagnani. This was about 1.5 miles above the town of Real de Santa Maria, near the limit of steam navigation. On 23 February they continued up the River Tuira by dugout canoe or piragua for about 30 miles to Boca de Cupé, arriving after dark. From here there was a 30 mile tramway to Cana. The first stage to Mount Kitchener, in the foothills, was by a short train drawn by a gasoline motor. For the next 12 miles to Paca the track wound up the mountain in tight curves which:

> were slowly and laboriously traversed by push car, a platform placed on trucks and pushed by men from behind. In descending, the cars,

> allowed to run by gravity, were rather insecurely controlled by coils of rope wound on the axles. (Goldman, 1912).

The final six miles from Paca was by a mule-drawn car. The journey from Panama took four days.

Seth E. Meek and Samuel F. Hildebrand (1912)

Two ichthyologists, Seth E. Meek and Samuel F. Hildebrand of the US Bureau of Fisheries, visited Panama in January to May 1911 and January to March 1912 during the "Biological Survey of the Canal Zone", conducted by the Smithsonian Institution, the Field Museum of Natural History, and the Bureau of Fisheries (Meek & Hildebrand, 1913; Heckadon-Moreno, 2004c). This resulted in a monograph on *The Fishes of the Fresh Waters of Panama* (Meek & Hildebrand, 1916). They travelled from Panama to Cana on 21 to 24 February 1912 in the company of Edward Goldman. They spent time collecting at Cana and gave a short description of their journey:

> The head of steam boat navigation is at El Real about 75 miles above its mouth. The head of tide is at Pinogana about 15 miles above El Real. ...
>
> The main stream was followed as far as Boca de Cupé, about 30 miles above the head of steamboat navigation. This distance was travelled in the native wooden dug-out. This stream has rather swift current above the head of tide, but there are no strong rapids, such as occur in the River Chagres. ....
>
> From Boca de Cupé the Darien Gold Mining Company has constructed a small tramway to its mines in the mountains at Cana, a distance of about 30 miles. This was followed and some of the small streams en route were fished. At Cana the small Rio Cana and the somewhat larger Rio Setiganti were visited and rather extensive collections made in each. An overland trip of about 10 miles was made from Cana to the Rio Grande another one of the upper tributaries of the Rio Tuyra. These small mountain streams are all very rocky and collecting is difficult. Most of the streams were clear and cool at the

time of our visit, but the Rio Cana was very turbid with a reddish sediment on its bed. (Meek & Hildebrand, 1916: 230).

3. Geology

The Espiritu Santo de Cana gold mine exploited high grade epithermal gold mineralization in a breccia pipe in a porphyry system. The geology and mine have been described by Restrepo (1886, 1888), Woakes (1895, 1896, 1899, 1923), Chibas (1898), Chase (1907), Low (1931), Oller (1933), Tower (1963), Tippett and Trever (1989) and Nelson (1995, 2006).

The deposit is located in the Pirre Massif, a NNE-trending rotated fault block in the transpressional East Panama Deformed Belt. The massif is bounded on the western side by a major N-S to NE-trending reverse fault, the Pirre or River Balsas Fault. Regional geological maps show the massif to be formed of an anticline of middle Eocene to Oligocene Darien Formation of tuff, agglomerate, chert and basalt in the lower part, and calcareous and siliceous mudstone, micritic calcarenite and volcaniclastic rocks in the upper part (Coates et al., 2004). The age of porphyry intrusion and mineralization is constrained to be late Oligocene or younger.

The deposit geology comprises andesite lavas and volcaniclastic rocks intruded by a circular hornblende-feldspar porphyry stock about 1,000-1,500 m in diameter and cross-cut by WNW-trending dykes (Nelson, 1995, 2006). Pervasive phyllic alteration with local potassic alteration is centered on the porphyry, with peripheral propylitic alteration. Drilling by Texasgulf and Freeport in the 1970s-1980s encountered deep porphyry-style mineralization including quartz veinlets with chalcopyrite. Other drill programs targeted epithermal gold mineralization at the North or El Rey Mine and San José hill, about which there is no geological information. Alluvial gold has been mined in the Cana Plateau, an intermontane basin east of the deposit.

Most of the historical gold production was from the South Mine or Espiritu Santo Breccia Pipe, which is 25 m diameter with 70° plunge to the SW, and was mined to a depth of 270 m with an average grade of 45-50 g/t Au (Woakes, 1899). The orebody on the 6th level was triangular in shape with sides of 45 x 36 x 24 m, elongated E-W (Chase, 1907). The breccia has angular rock fragments cemented by pyrite, sphalerite, galena, visible gold, quartz and calcite. Woakes (1896) described the country rock as grey feldspar porphyry with strong fracturing with barren pyrite. The Espiritu Santo lode was described as "at least three types of conglomerate" (breccia) with angular to rounded clasts of porphyry cemented by calcite, small acicular quartz crystals, pyrite, sphalerite and galena. Gold was coarse, though rarely visible, on sulfides and gangue. The bullion assayed 920 to 930 fine (92-93% Au)(Woakes, 1896).

The Espiritu Santo deposit is interpreted as high grade Au-bearing, intermediate sulphidation epithermal mineralization with base metals in a phreatic or phreatomagmatic breccia pipe related to the Cana porphyry Cu-Au system. Gold mineralization at the North and San José Mines is probably also epithermal. The porphyry stock and mineralization crosscut volcanic rocks of the Darien Formation, indicating an age of late Oligocene or younger.

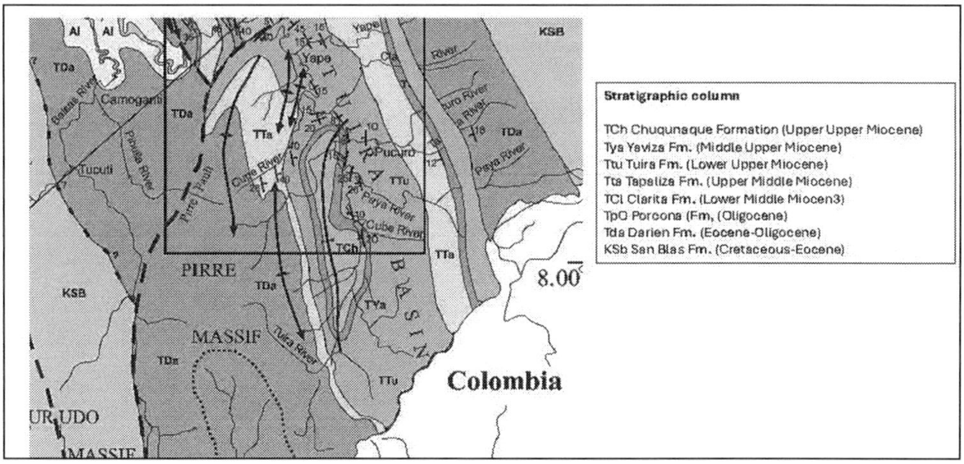

Above: The regional geological setting of the Cana gold deposit (Coates et al., 2004).

Below: The geology and mines of the Cana deposit (map by S. Redwood based on Nelson, 2006).

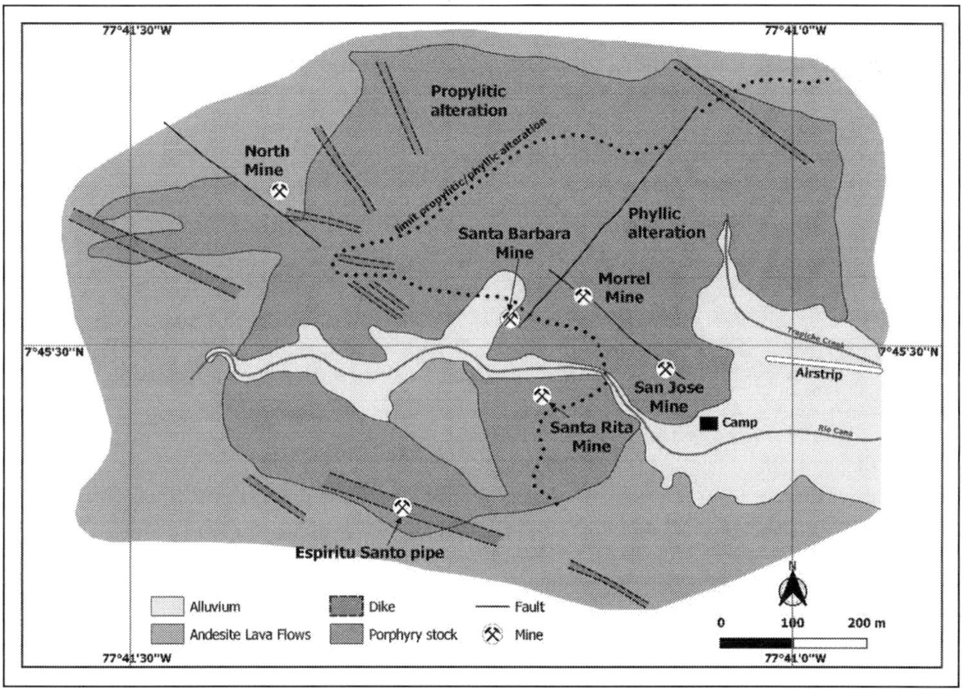

4. Prehistory

Despite the popular belief that the Darien is pristine rain forest, there is botanical and historical evidence for significant human modification of the forest for agriculture, followed by regrowth. Dolores Piperno of the Smithsonian Tropical Research Institute, Panama (STRI) analyzed fossil pollen in sediment cores collected from Lake Wodehouse and Cana Swamps 1 and 2 in 1988 (Bush & Colinvaux, 1994; Piperno, 1994). Lake Wodehouse is a large flat area in the drainage of the River Tuira about 15 km east of Cana with a small permanent body of water of about 300 m by 70 m in size that expands to about 9 km^2 during the rainy season. (Lake Wodehouse was an informal name given by Paul Colinvaux after the writer P.G. Wodehouse, his favorite author, one of whose books accompanied the expedition.)[5] A 10.7 m long soft sediment core was recovered, ending in gravel and dating to 3,910 ± 80 BP. Cana Swamp 1 near the Cana gold mine was also cored. A 10 m sequence dating to 3,740 ± 110 BP was retrieved, but there were too few phytoliths and the study focused on Lake Wodehouse. There are no archaeological records for the region, the only cultural evidence of past human activity being manos and metates that Piperno found lying in the rubble of the airstrip construction at Cana. Palaeobotanical data from the Lake Wodehouse core indicate a minimum date of 4,000 BP for the beginning of prolonged human habitation and prehistoric agriculture. Early systems of slash-and-burn cultivation included maize (Zea mays) as a primary food plant and appear to have resulted in significant forest modification. An episode of forest regeneration shortly after 350 BP (about 1600 AD) indicates a decline in the indigenous population inferred to be as a result of European influence and suggests that the extant forest near the site is recent regrowth after long-term prehistoric occupation and agriculture. Local forest recovery around Lake Wodehouse took less than 350 years, little more than one or two tree lifetimes, to attain the present richness of forest that surrounds the site.

Bray, Cooke & Redwood (2021 and references therein) examined the myth of the "Darien Gap" with impenetrable rainforests. It dates from colonial times when it was used to embellish the bravery of Spanish soldiers, such as Vasco Núñez de Balboa, when he "discovered" the Pacific Ocean. It is, in fact, a misleading cliché. The early Spanish chroniclers describe a landscape that was a patchwork of agricultural fields, second growth, and riverine and montane forests. It was well-populated with warring-cum-trading communities controlled by powerful caciques, and linked together by coastal and riverine canoe routes, and by overland trails along which a variety of merchandise was transported. This agrees with the evidence from the sediment cores for slash-and-burn clearing for the cultivation of maize. Supporting archaeological evidence is provided by Louis Catat (1889), a French engineer who reported Pre-Columbian cultural materials near the Cana gold mine. In summary, several lines of evidence indicate that there were Pre-Columbian settlements in the Darien and extensive cultural interchange, with no evidence of a barrier.

5. Spanish Gold Mining

Early Spanish Gold Mining in Darien (1513-1559) and Veraguas (1524-1620)

Gold was first mined by the Spaniards in the northern Darien goldfield in the early 16th century near the first towns of Santa María la Antigua del Darien (1514-1524, now in Colombia) and Acla (1520-1559) until the Spanish focus moved to the Pacific and founded Panama in 1519 (Mena García, 2011; Blas Aritio, 2012; Redwood, 2020). From here the conquest moved west to Veraguas where numerous goldfields were discovered over the next century between 1524 and 1620. The Spanish did not return to the Darien for 150 years. The Cana gold mine in the interior of eastern Darien was not discovered until the late colonial period and was the last significant mine to be found in Panama.

The Discovery of Santa Maria (1665) and Cana (1689)

The placer deposits in the Tucuti district in the Darien goldfields were first worked in about 1665 when the fort and mining town of El Real de Santa Maria was founded on the River Tuira, 40 km north of Cana. The Espiritu Santo de Cana mine was discovered by Capitán Antonio de Landeche in about 1675 and was in operation by 1680 as shown by Spanish and English documents. Santa Maria was raided by English pirates in 1680 and 1684, and the Cana mine was raided by pirates in 1702, 1712 and 1724. The Cana mine was abandoned in 1727 after almost 50 years of production. (Restrepo, 1886; Joyce, 1934).

The Cana gold mine was discovered in about 1675, shortly before it is first described in an official document of 1680:

> The President of Panama gives an account of having promoted the discovery of gold minerals in the Province of Darien and of having begun their exploitation with great hope of their wealth by issuing documents of the matter.[6]

Captain Bartholomew Sharp's Raid on Santa Maria (1680)

News of the gold discovery travelled fast. Cuna (Guna) Indians informed a large group of English buccaneers on seven ships anchored at Golden Island in San Blas about the gold mine in March 1680. Led by Captain Bartholomew Sharp, they planned to raid Santa Maria to obtain the gold that was brought down from the Cana mine in the first Buccaneer Expedition to the South Sea of 1680-1682. The date corroborates the Spanish document cited above for the discovery of the Cana mine a few years before 1680. A group of 331 English pirates set out from Golden Island to cross the isthmus on 5 April 1680, leaving two captains to guard the ships. The crews marched under their respective commanders with distinguishing flags. They were led by Captain Bartholomew Sharp (c1650-1702) and followed by Captain Richard Sawkins, Captain Peter Harris, Captain John Coxon and Captain Edmund Cook. Chiefs Andres and Antonio were in charge of the Darien Indians, and there were even a few Misquito Indians from Cape Gracias a Dios, allies of the English.

A detailed log was kept by Basil Ringrose (c1653-1686) that was published as the final section of the English translation of the French buccaneer John Esquemeling's (c1645-1707) *The History of the Buccaneers of America* in 1685, first published in Dutch in 1678 under the name of Alexandre Olivier Exquemelin, a best-selling book that romanticized pirates and is still in print today (Esquemeling, 1685/1911, Part IV, p. 273-502; Masefield, 1906, p. 232-244; Anderson, 1911). The group included the four so-called "literary pirates" who wrote books about their adventures. As well as the ship's doctor Ringrose, they were the buccaneer navigator and naturalist Captain William Dampier (1651-1715) who wrote a famous book *A New Voyage Round the World* (Dampier 1678/1998), the buccaneer surgeon

Lionel Wafer who wrote a description of the isthmus and the Cuna (Guna) Indians (Wafer, 1695 / Joyce, 1934), and Bartholomew Sharp who kept a journal (Sharp, 1684). As Masefield (1907, p. 234) wrote, "It is not often that three historians of such supreme merit as Dampier, Wafer, and Ringrose, are associated in a collaboration so charming, as a piratical raid."

The hard isthmian crossing took nine days. After a day's rest, a force of 327 Englishmen and 50 Indians set off in 68 canoes to paddle down the River Tuira on 14 April and landed near Santa Maria well after midnight and some two hours before dawn. The pirates attacked the fort at daybreak on 15 April.

> The fort was formed simply of palisades, some of which the English took down and entered without difficulty. The Spaniards surrendered without much opposition; nevertheless, twenty-six were killed and sixteen wounded. (Anderson, 1911).

The pirate rendezvous of Golden Island (Suletupu), San Blas.
(photo S. Redwood, 1993).

The Governor, the priest and the chief citizens escaped down the river. The Indians killed 50 Spanish prisoners before the pirates put a stop to the massacre. They only found about twenty pounds of gold, having just missed three hundredweight of gold that was shipped to Panama only three days before. The buccaneers burned the town and departed down the river on 17 April 1680. They spent the next two years raiding the Gulf of Panama and the South Sea from Nicaragua to Chile.

Basil Ringrose's account was published in Esquemeling (1685/1911):

> This golden Treasure cometh down another branch of this River unto Santa Maria, from the neighboring Mountains, where are thought to

Bartholomew Sharp's Adventures, 1684.

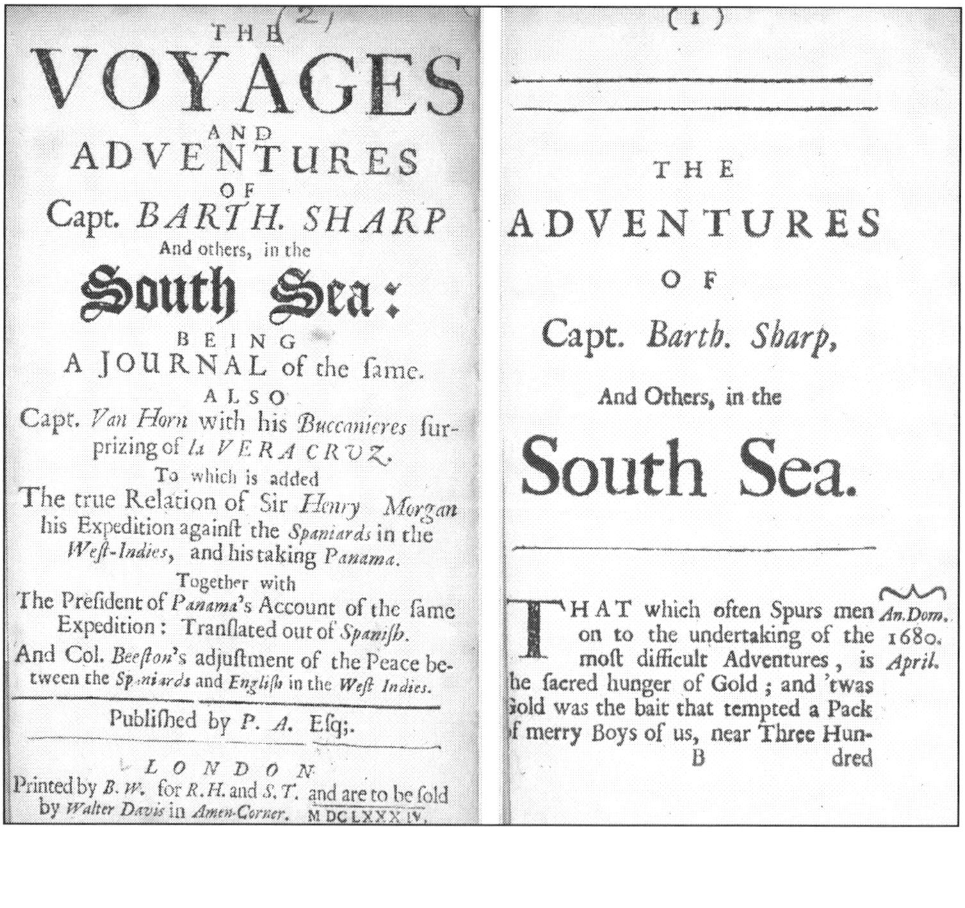

(2)

THE VOYAGES AND ADVENTURES OF Capt. *BARTH. SHARP* And others, in the South Sea: BEING A JOURNAL of the ſame. ALSO Capt. *Van Horn* with his *Buccanieres* ſurprizing of *la VERA CRUZ*. To which is added The true Relation of Sir *Henry Morgan* his Expedition againſt the *Spaniards* in the *Weſt-Indies*, and his taking *Panama*. Together with The Preſident of *Panama*'s Account of the ſame Expedition: Tranſlated out of *Spaniſh*. And Col. *Beeſton*'s adjuſtment of the Peace between the *Spaniards* and *Engliſh* in the *Weſt Indies*.

Publiſhed by *P. A.* Eſq;.

LONDON: Printed by *B. W.* for *R. H.* and *S. T.* and are to be ſold by *Walter Davis* in *Amen-Corner*. MDCLXXXIV.

(1)

THE ADVENTURES OF Capt. *Barth. Sharp*, And Others, in the South Sea.

THAT which often Spurs men on to the undertaking of the moſt difficult Adventures, is the ſacred hunger of Gold; and 'twas Gold was the bait that tempted a Pack of merry Boys of us, near Three Hundred

An. Dom. 1680. *April.*

B

be the richest Mines of the Indies, or, at least, of all these parts of the Western World. ... But as bad a place as it was, our fortune was much worse. For we came only three days too late, or else we had met with three hundred weight of Gold, which was carried thence to Panama in a Bark [boat], that is sent from thence twice or thrice every year, to fetch what Gold is brought to Santa Maria from the Mountains.

Bartholomew Sharp's account was as follows (Sharp, 1684, p. 9):

Thurfd. About fix this morning we attacqued the place and carried it with little difficulty, it being a Stockadoe Fort, and a fmall Town of Thatched Houfes. This Fort oi Santa Maria was kept by the Spaniards for the conveniency of gathering Duft-Gold, which the River affords plenty of, and the poor Natives are the Drudges to gather it for them. We defigned to make no further progrefs, being told there was a fuffi-cient quantity of Gold-duft at this place to enrich us all, but the wary Spaniards had carried it away two days before, as thinking a convey-ance of their own contriving fafer than ours, though not fo well guard-ed as it would have been by us.

William Dampier described the raid very briefly because Ringrose had described it. He wrote:

The first Expedition was to Portobello, which being accomplished, it .was resolved to march by Land over the Isthmus of Darien, upon some new Adventures in the South Seas. Accordingly on the 5th of April 1680, we went ashore on the Isthmus near Golden Island, one of the Samballoes [San Blas Islands], to the Number of between three and four hundred Men, carrying with us such Provisions as were necessary, and Toys with which to gratify the Wild Indians, through whose Country we were to pass. In about nine Days' march we arrived at Santa Maria and took it, and after a Stay there of about three Days, we went on to the South Sea Coast.... (Dampier, 1697/1998, p. 4).

Lionel Wafer also described it briefly (Joyce, 1934, p. 3):

> For in short, having muster'd up our Forces at Golden-Island, and landed on the Isthmus, we march'd over Land, and took Santa Maria; and made those Excursions into the S Seas, which Mr. Ringrose relates in the 4th part of *The History of the Buccaniers*.

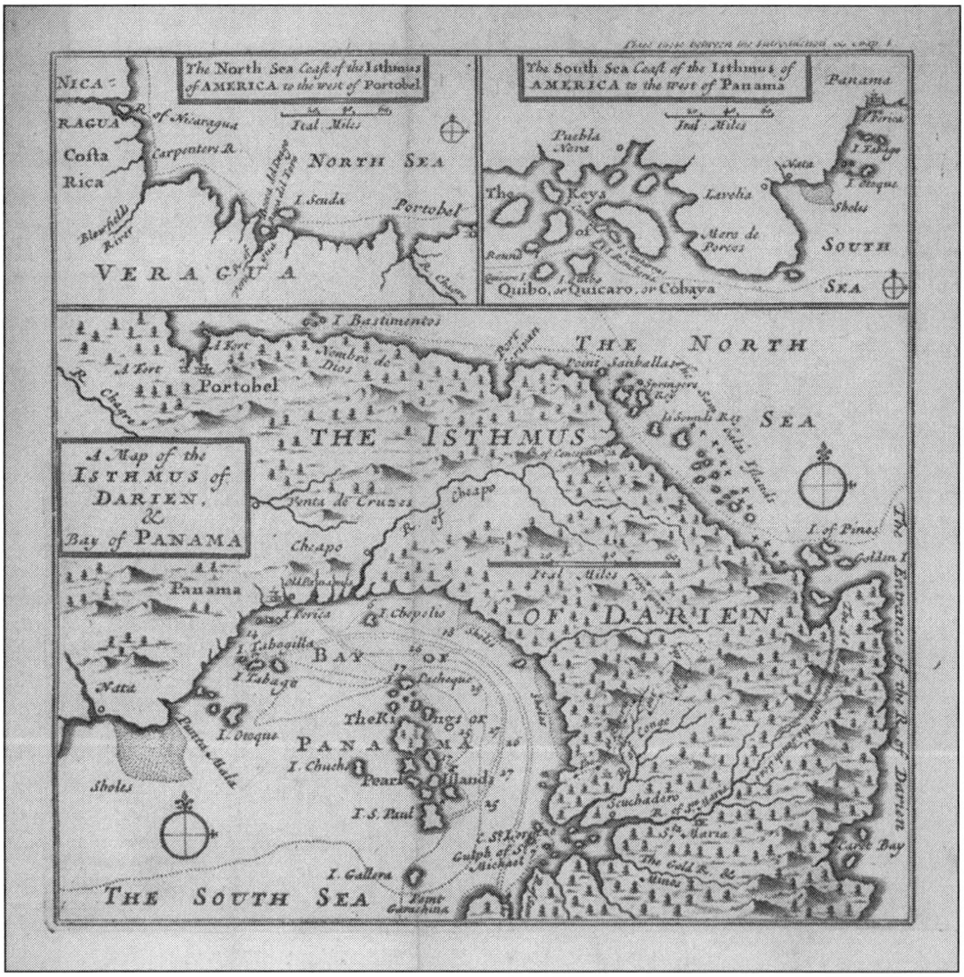

Map of the Isthmus of Darien showing Golden Island, Santa Maria and 'The Gold River and Mines' (Dampier, 1697).

After a year of the expedition, a group of 47 buccaneers separated from Captain Sharp on 17 April 1681 at Isla de Plata in Ecuador and sailed north in a long boat and canoes for 12 days to return to the Gulf of San Miguel, followed by a grueling 23 day crossing of the isthmus back to their ships in the Caribbean. This group included William Dampier, Lionel Wafer who was injured at the River Congo and stayed behind to live with the Cuna (Guna) Indians for four months until he had recovered, and John Cook, later a buccaneer captain. Captain Sharp continued his piracy in the South Sea and returned by way of the Strait of Magellan to the West Indies in January 1682. On his return to England, he was arrested for piracy at the instance of the Spanish ambassador, tried and acquitted.

Santa Maria Seized by Peter Harris (1684)

El Real de Santa Maria was captured again by English pirates in 1684 (Restrepo, 1886; Joyce, 1934). This expedition was led by the young Captain Peter Harris, nephew of the Peter Harris of the 1680 expedition in which he was killed in Panama. Harris crossed from Golden Island in June 1684 and seized Santa Maria. The Spaniards were driven out or fled and Harris obtained 120 pounds of gold, including a nugget "as big as a hen's egg". He then went to the Gulf of San Miguel where he seized a barque and joined William Dampier in Ecuador six months later (Joyce, 1934; Dampier, 1697/1998, p. 85).

According to Cullen (1853, p. 47-48), the gold mines of Darien were closed by Royal Decree in 1865 which reads as follows:

> Year 1685. Royal decree. March 12th. - That the President of Panama break up and destroy the mines of gold that exist in the vicinity of the rivers of the province of Darien, because the coveting of them has induced the pirates to undertake the transit from the sea of the north to the sea of the south by those rivers, to the prejudice of the public cause - and that the Viceroy of Peru cooperate in it.[7]

However, it appears that it was not carried out and both mining and pirate raids continued, as Cullen narrates:

> GOLD MINES ON THE UPPER TUYRA - WHY SHUT UP. On Cana and some other branches of the upper course of the Tuyra, the old Spaniards carried on gold mining very successfully; but the mines were closed by order of the King of Spain, in consequence of their having attracted the Buccaneers. (Cullen, 1853, p. 58).

William Dampier's Description of Santa Maria (1694)

William Dampier (1651-1715) was the greatest explorer of the 17th century and the most important explorer between Sir Francis Drake in the 16th century and Captain James Cook in the 18th century. He circumnavigated the world three times as a buccaneer and also made an expedition to Australia in command of a Royal Navy ship, where he is regarded as Australia's first naturalist. He rescued a former shipmate, Alexander Selkirk (1676-1721), from four years as a castaway (1704-1709) on the Juan Fernandez Islands, the inspiration for Daniel Defoe's *Robinson Crusoe* (1719).

Dampier described the gold placers of Santa Maria based on Peter Harris's raid of 1684 (Dampier, 1697/1998, p. 102-103):

> About six Leagues from the River's Mouth, on the South-side of it, the Spaniards, about 20 years ago, upon their first Discovery of the Gold Mines here, built the Town of Santa Maria. This Town was taken by Capt. Coxon, Capt. Harris and Capt. Sharp at their entrance into the South Seas, it being then newly built. Since that time it has grown considerable. For when Captain Harris, the Nephew of the former, took it, he found in abundance all sorts of Tradesemen, with a great deal of Flour and Wine, and abundance of Iron Crosses and Pickaxes. These were Instruments for the Slaves to work in the Gold Mines. For, besides what Gold and Sand they take up together, they often find great lumps wedged between the Rocks, as if they naturally grew there. I have seen a Lump as big as a Hen's Egg, brought by Captain Harris (who took 120 pound there), and he told me that there were Lumps a

Detail of a 1733 map by Henry Popple showing 'The Gold River and Mines' and 'Santa Maria'.[8]

great deal bigger. But these they were forced to beat in pieces, so that they might divide them. These Lumps are not so solid, but that they have Crevices and Pores full of Earth and Dust.

This Town is not far from the Mines, where the Spaniards keep a great many Slaves to work in the dry time of the year. But in the rainy Season when the Rivers overflow, they cannot work so well. Yet the Mines are so high the mountains that as the Rivers soon rise, so they are soon down again; and presently after the Rain is the best searching for Gold in the Sands, for the violent Rains do wash down the Gold into the Rivers, where much of it settles to the bottom and remains. Then the Native Indians who live hereabouts get most, and of them the Spaniards buy more Gold than their Slaves get by working. I have been told that they get the value of 5 Shillings a day. Most of the Spaniards withdraw with their Slaves during the wet season to Panama.

William Dampier (circa 1650-1717) was the greatest explorer of the 17th century. Painted holding his book, circa 1697-1698. © National Portrait Gallery, London.[9]

William Paterson's Description of Cana (1700): Scots Colonial Dreams

The first description of the Cana mine was actually written by a Scotsman in 1700, William Paterson (1658-1719), the promoter of the ill-fated Scots colony of Caledonia (1698-1700), Darien, located 120 km north of the mine. Paterson was a Scottish businessman who had been a merchant in the West Indies and founded the world's first central bank, the Bank of England, in 1694 (Forrester, 2004). "The Company of Scotland Trading to Africa and the Indies" was founded in 1695 and set up the colony of Caledonia in 1698, but it was short-lived and was abandoned in 1700 after being forced to capitulate by the Spanish governor (Prebble, 1968).

Paterson wrote a long description of the Cana mine in "Mr. Paterson's Last Plan for the Darien Company,"[10] a prospectus with a detailed description of the colony written in 1700 after the failure of the First Expedition (Dalrymple, 1778). He proposed a new joint stock company to run the colony with capital of £2,000,000, in which King William of England would have joint dominion of the colony. Four-fifths of the capital would be offered to English shareholders, with the remaining one-fifth or £400,000 being the capital that Scotland had already raised. The total capital was five times the capital of the original Scottish company and double the size of the Bank of England (Forrester, 2004, p. 295-297). The prospectus was reproduced in 1778 by Sir John Dalrymple of Cousland, 4th Baronet, FRSE, FSA Scot (1726-1810), in Volume III of his history of the British Isles, *Memoirs of Great Britain and Ireland* (Dalrymple, 1778), and contains a description of the Cana gold mine:

> But the gold mines of this country, and its two paffages from the north to the fouth fea, are above all other things the moft valuable. Gold may be gotten in greater or leffer quantities in very many places of the Ifthmus: But fince the late difcoveries no mines are lookt after that will not yield at leaft half an ounce of gold per diem to the labourer: But they often yield a much greater quantity, even to half a mark or four ounc-

es per day; and many thoufand negroes or others might find conftant employment at this rate.

In former times, the Spaniards had no mines in this country, but only a place not far from St. Maria, where gold in grain was found to the value of about a Caftilian, or the fixth part of an ounce per day; and fometimes they had poffeffion of this, and at other times not, according as there was peace and war with the Indians: But, about 16 years ago, from the difcoveries of the natives to fome privateers, the three mines, called Sabalas, Archietee, and Talieque were found; and after the peace that then enfued, the Spaniards were permitted to work them. The mines of Talieque and Archietee lye each about a day's journey from Sta. Maria, and between thefe two there is only the diftance of three leagues. From Archietee the Sabalas is five leagues, and from thence to the great mine of Cana there is about three days journey. This mine of Cana, with another of the like nature, about a day's journey nearer to the Cape Tiberoon, were both difcovered by the natives about feven years ago. Thefe mines confifted not only of gold in grain, in common with others of the country, but here was alfo found great quantityes in vein and in ftone. Gold in grain is fuch as they gather from the wafhings of the floods; gold in vein is where it is found in veins like other metals; and gold in ftone is where gold is found intermixt in the rocks or quarryes of ftone. After the difcovery of thefe mines, the Indians continued to get great quantities of gold even by their lazy and untoward way of working, until the Spaniards, who were then at peace with the natives, came to an agreement with them to fend a number of negroes to work on certain conditions: But the Spaniards, having gotten in their finger, foon broke conditions with and difobliged the Indians, and this occafioned the prefent war between them; and for the recovery of this mine it was that the natives brought affiftance at our firft arrival. And although the mine does, without all doubt, belong to the natives, and we might juftly enough have lent them affiftance to recover their right; yet we declined it, leaft we fhould thereby happen to give the leaft fhaddow or colour of offence. The mine of Cana continues ftill to be wrought by 1000 negroes, befides others: But the Span-

William Paterson (1658-1719).[11]

iards have done what they can to ftop up and ftifle that next to Cape Tiburoon; becaufe of its nearnefs to the north fea, and confequently of its vicinity to powerful neighbours. So far as I can learn, this mine of

> Cana lyes about 15 leagues from your harbour of Caledonia, towards the fouth-eaft.

According to Paterson, the first Darien gold mines had been discovered 16 years previously, which is about 1683, and the Cana mine was discovered about seven years previously in about 1692, but these dates are later than those indicated by other accounts. The Scots did not visit the Cana mine, but Paterson evidently got his information firsthand from the Cuna (Guna) Indians who visited Caledonia in 1698-1699 to ask the Scots for help to recover the mines from the Spanish. The Scots declined to help in order to avoid conflict with the Spaniards. The mine was being worked at the time by "1,000 negro slaves" and others. There was another similar gold mine closer to Cape Tiburon which the Spaniards had hidden to protect it.

The Scots colony did not carry out any gold mining but Paterson made frequent reference to the Cana and other Darien gold mines to promote the colony. He referred to the pirate accounts of the mines to legitimize the legal status of the colony as tierra franca or open land. Indeed, the pirate accounts by Wafer and Dampier gave Paterson the idea for the location of the colony. As narrated here, Paterson wrote the first description of the Cana mine. However, he did not visit the mine out of respect for Spanish sovereignty, for the Scots came in peace and not as raiders. One ex-Caledonian, Robert Allen, later took part in a pirate raid on the mine. It appears that the Cana mine influenced the location of the Caledonia colony.

Nathaniel Davis's Raid (1702)

Nathaniel Davis raided the mine in 1702 with 482 English pirates and 100 Cunas (Gunas) of Cacique Pedro. He reported that the town of Cana, which they burned, had about 900 houses and one church. *Davis' Expedition to the Gold Mine* (1704) account was the first eye-witness description of the mine and was reproduced by Joyce (1934) in the Hakluyt Society edition of Lionel Wafer's description of the Darien.

> On the first of September [1702] we sent out a company of our Men, with Spaniards and Negroes, to wash the Gold from the Oare [ore]; the Mine is on the side of a great Hill, above Thirty Yards [27.4 m] deep, and several Caves [tunnels] run into the Hill farther than any one would venture to go; the Oare they dig out of it is a sort of a mixture of Rock, which after it is dug out of the Mine is brought to the Mill, which grinds it small, then 'tis washed, made up into the form of Bricks and lodged in Houses built at the Mine for that purpose, over which a Guard is set, with a Captain and a Governour to see that the King is not cheated. After it has lain some short time in those Houses, then it is washed a second time, and so cleared of the Dross or some rocky Part wherewith it's intermixed till there remains pure Gold. They make a great quantity of Gold every Day they work; we made five pound weight and nine Ounces [about 2.8 kg] in less than a Day. Everything here sells at an Extravagant Rate, as a pound of Sugar at fifteen shillings, and so proportionately for all things else; ....
>
> Next day our Scouts brought in more Spaniards and Negroes than we already had in our Power, some of whom we sent with a Guard to the Mine to wash the Ore, of which they made six pounds weight.
>
> We sent twenty-four Negroes on the third into the Mine, who brought eight pounds of Gold.... We made however fourteen pound weight of Gold on the 4th, as we did sixteen on the next; [they got] Mules ready to carry our sick men [and left on the 7th] .... Then we fired the Town wherein I guess there might be about nine hundred houses, but one Church.

The pirates extracted a total of 49 lbs. 9 oz. (about 24.6 kg) of gold in four days.

Robert Allen, a Scots Deserter Cum Pirate (1702)

Another member of Davis's 1702 pirate expedition was Robert Allen, a Scots merchant who had deserted from the Scottish Darien colony in 1698 and went to Jamaica. He returned to Darien in 1701 to explore for logwood, where he was the only survivor of an Indian massacre and then lived with friendly Indians until joining the 1702 pirate raiding party to Cana (Alsop, 1986; Orr, 2018, p. 19-21). His report shows the size of the expedition which numbered 600 men (Alsop, 1986).

> Being at length informed by them that nine English Sloops were on that Coast I gott on board of these and found about 600 men of Her Maty's [Majesty's] Subjects that had commissiones from the Governour of Jamaica and were designed on an expedition against Her Maty's Enemies in the Spanish West Indies. I Joyned my self to them and prevailed with 60 Indians to accompany us. 560 of us marcht from ye mouth of the River Chocó and took the town and mines of Cana and returned to our sloops in less than a month with good Booty.

The pirates raided from the Caribbean by an unusual route from the east by sailing south up the River Atrato (Chocó) and west up the River Salaqui, then crossing the Quia Mountains at Altos de Quia to reach Cana. He later took part in a pirate raid to Antioquia but was captured, taken to Cartagena and eventually Quito.

> Afterwards 210 of us went out again on a designe to have taken the town & mines of Antiochia and in our way took ye mines of Bevera [not identified]. But after a moneths travill many dificulties, severall Incounters & ambuscadoes our company was reduced to about 150 and being Surrounded with Spaniards & Indians and in want of all necessarys we surrendered upon promise of quarter & safe conduct to Carthagena.

Charles Tibón's Raids (1712, 1724)

A Spanish map of Darien dated 1774, described by de Puydt (1868), shows the village of Santa Cruz de Cana "which, in the year 1712, was set fire to, amidst the slaughter of the inhabitants, by 80 Frenchmen and 300 of the Gulf Indians, under the command of Charles Tibou [Charles Tibón]."[12]

In 1724, a group of pirates led by the Frenchman, Charles Tibón, "put to the sword the whole garrison of Chepigana fort and destroyed the galleries and tunnels of the mine which had to stop production."[13]

Mine Collapse, Cuna (Guna) Revolt and Mine Closure (1727)

The mine was abandoned in 1727 due to the collapse of the shaft which resulted in the death of two miners. Other factors were the frequent pirate raids, as well as the violent indigenous insurrection in the Darien in 1727, which was not pacified until 1778. The mine was not worked again in the colonial period.

In 1727 a violent Cuna (Guna) rebellion took place, led by a mestizo Luis Garcia with 200 Cuna (Guna) Indians armed with 70 guns. They attacked the Cana mine and burned El Real including the church, Yaviza, Chepigana and other villages, and killed 28 Spaniards and several women, children and slaves, and a Dominican priest Brother Ambrosio Gómez. Garcia called himself the Liberator of Darien and his plan was to invade Portobello, Panama and Cartagena. The rebellion was bloodily put down, Garcia was captured and sentenced to death, but this was commuted to exile in Peru. (Castillero Calvo, 2019a, 2019c).

Although the mine was closed, the Spaniards maintained permanent military posts in the mid- to late-18th century at El Real de Minas de Cana, el Real de Santa Maria with a wooden palisade fort, and at Chepigana, as a result of geopolitical interest in Darien (Castillero-Calvo, 2019b).

Restrepo (1886) wrote of the mine closure:

> It will doubtless surprise the reader to know that so rich a mine should have been left abandoned for a century and a half. The reason, however, is a very simple one. After the frequent attacks of the pirates who took possession of Cana in 1684, 1702, 1712 and 1724, occurred the general uprising of the Indians instigated by a bold half breed named Luis Garcia. This rebellion, which took place at the same time as the caving in of the Espiritu Santo mine, was so terrible that all the well-to-do families deserted Darien and fled to Panama, Cartagena and other provinces, "not being able to live"

says Ariza, "in the continual fear of an attack and of falling victims to the cruelty of the savages." Only the most miserable inhabitants were left in the province. Peace was not wholly restored until 1778; but then Darien had become almost a wilderness, and its population which, in 1727, had amounted to 20,000 was reduced, according to Ariza, to about one thousand souls.

Darien Province (1774)

According to de Puydt's description, the 1774 map shows there were nine (Spanish) villages in Darien with a total population of 1,145 (de Puydt, 1868). They were (de Puydt's spelling used for place names): 1) Yavisa, fortified, with 170 inhabitants, on the River Chuqunaque; 2) Santa Maria la Reale, fortified, with 180 inhabitants, at the mouth of the River Pirré; 3) Cana, fortified, with 26 inhabitants, on the River Cana; 4) Chepigana, fortified, 26 inhabitants; 5) Tichiche, 117 inhabitants; 6) Pinogana, 180 inhabitants; 7) Molineca, 110 inhabitants; 8) Tucutí, 168 inhabitants; and 9) La Marea on the River Tayecua with "much gold" and 168 inhabitants.

The map also shows the ten Indian Villages named Congo, Balsas, Acauti, Paya, Yavisá, Sambú, Pirré, Matumaganti, Tapunaca, and Tupisa.

There were 15 gold mines in the province at Troncoso, Sábalos, Tayecua, Nusaganti, Arquiati, Nususunaqui, Acuasiscuati, Bagre, Marea, Balsas, Cana or Espiritú Santo, River de Playon, Sucubti, Cuque (mouths of the River Atrato, now in Colombia), and Mali, situated on an arm of the Pucro (de Puydt, 1868).

Don Andrés de Ariza's Reports (1774, 1778, 1780)

Don Andres de Ariza, Governor of the Province of Darien from 1774-1794, wrote a long report addressed to the Viceroy Guiror in 1774 (Ariza, 1774; Restrepo, 1886; Rodriguez Hernandez, 2016).[14] This is the most detailed description of the Espiritu Santo mine, and despite being closed for 50 years, he interviewed people who had worked in the mine, and particularly its master and mover, Pedro Oramunio.

Restrepo (1886) praised Ariza highly.

In our national history, a brilliant page should be consecrated to the memory of Lt. Colonel Don Andres de Ariza. His mind teemed with grand ideas, which he attempted to carry out, and that he did not succeed was due to the fact that the means at his disposal were very limited. He did succeed in restoring peace to Darien, and constructed fortifications to resist the attacks of the savage Indians. He made accurate maps of the province, and, with the approval of the King of Spain, opened a road which brought the two oceans that wash the shores of Darien, into rapid communication with each other. It was he, in short, who made it possible for the Spanish government to operate the rich mines of Cana. In 1777, he went to Cana, "to build," he says, "a fort or strong-house, but in the hope that it will be used more for the safe keeping and storing of the gold extracted from the magnificent mines, [because it always has seemed to me that the Government will not allow them to remain deserted long,] than to resist the attack of the neighboring Indians." The wealth of the mines of Santa Cruz de Cana has been preserved in tradition; all the explorers who have visited Darien speak of these mines with admiration, but the road that led to them was overgrown with virgin forests, and has been finally forgotten.

Ariza wrote in 1774:

The hills of Santa Cruz de Cana, and others more to the north, are so abundant in ore that to speak of their magnificence seems more like hyperbolical exaggeration than actual reality. But as they are known not only by the stories of men of veracity, but also because there are at the present time various persons living who have actually worked in the famous Espiritu Santo mine, from whom exact information has been obtained, no doubt can remain of its vast wealth and positive existence in that place.

The Espiritu Santo mine was in operation down to the year 1727. It was well equipped; the actual vein of gold it contained was very extensive and of such fine metal that it exceeded 22 carats. The direction of the vein was inclined, or nearly perpendicular, so that it could be worked very profitably in spite of the great cost; it had five ladders of 12 or 15 steps each, for the use of the operators in entering the mine. There were also four wells in which the water that filtered through the earth was collected until it could be thrown into the river.

More than 200 men were employed in it, who worked alternately, night and day, and in separate operations; they followed the main vein and several branches, which at times were found to be very rich, taking out of the mine earth and pieces of flat stones containing gold, which were sent to be washed. The miners were so placed, that the ore, &c., was passed from hand to hand until· the bottom of the shaft was reached, where, by means of a machine operated by two men in a wheel, it was drawn to the surface, where it was taken to the river to be washed.

The mine had four levels, but the lowest was the largest and widest. A great many miners worked in it without in the least interfering with each other. At intervals pillars of earth were left for support, and in some parts the excavations were lined with heavy timber as a protection against a cave-in. Unfortunately this precaution was not taken in the shaft where the two miners worked the machine, and the gravity of the soil overcoming its coherence it caved in, from all four sides, burying-beyond the possibility of human aid-the two poor follows who were employed in sending the soil up to the surface of the mine.

This bad omen, together with the losses suffered through the piracies of foreigners and the Indian attacks, dismayed both owners and miners, who, fearing that the same accident might occur in other parts of the mine, became panic-stricken, (and are still believed to be so) ascended the ladders, which were available, resolved never to return to work. For this reason the owners of the mine gave it up, and went to Panama, nor did they return to this province.

Almost all the operatives of this mine were free blacks; their wages, which they received every Saturday, was a small dish of earth for each day of the week; and if the owners thought the soil was poor in gold instead of six of them they would give seven, which after washing would yield not less than 16 to 20 castellanos, and sometimes would yield 40 or 50.

Nor were the wages of these unfortunate people their only source of riches, for they stole as well, and each time they went out of the mine, (if by chance they were allowed to do so), and also when they carried their food in, they took the opportunity to go to the richest part of the vein and from this or from one of its branches they stole the choicest ore; there was always abundance for both robbed and robber, the metal at that time being sold by the bottle full. " Master Pedro Oramunio, who was in the mine at the time the shaft caved in, relates that whenever the workmen were in urgent need, as they knew well where the vein was located, (though this was usually kept secret), they would go to it, fill their pockets or a bag with the earth, and when they thought the guards were asleep, would leave with their loaded bag, from which they would extract at least 3 to 5 lbs. of gold; such richness seems incredible, but numerous persons vouch for its truth.

As a confirmation of the magnificence of Santa Cruz de Cana, it is said that the negroes, mulattoes and "zambos", (children of Indians by negros or vice versa) who worked in the mine, when they went to balls with their sweethearts, would to gain the favor of their ladies sprinkle their heads and even their feet with the gold dust that they carried loose in their pockets.

It is also an undisputed fact, that a slave of Don Antonio de Sosa (Don Antonio Arguelles, Don Diego de la Guardia, Don Juan de Leon, Don Diego Mogica and Don Antonio de Sosa were the last owners of the mine) found in the mine a deposit, or as the miners say, a pocket of gold, such as were frequently met with. But this being of unusual size, he went to his master with the news, asking for a reward; this his master granted him, setting him and his wife free, giving them a house in Panama, a little farm and some money besides to stock them with. Several persons assert that, according to the report of Señor Sosa himself, this pocket yielded 60,000 castellanos of gold (600 pounds); others say 50,000, but Oramunio declares that it gave from 16 to 20 thousand castellanos, (160 to 200 lbs.); any one of which estimates

> indicates a magnificent yield.
>
> It is a common saying that the mines of Espiritu Santo was supported by pillars of gold; this expression owed its origin to the fact that in the mine pillars of earth were left here and there to prevent its caving in. All parts of the mine being rich in gold and the pillars being very convenient to those who worked in it, every day a little earth was stolen from them, at least enough to fill the pockets ; the consequence being that the pillars became in time so diminished in size that it was necessary to cover them with heavy planking to prevent caving in.
>
> Don Juan de lbiricu at last undertook to operate the mine of Espiritu Santo putting it in running order, and though he had but few slaves extracted considerable metal; but having neglected to secure a title to the mine from the tribunal of the Royal Treasury, he was prevented from continuing the work by a decree of the Court, and in this strait the mine was again abandoned.

Restrepo (1886) also quoted from two other important manuscripts by Ariza in the collection of Señor Don José M. Quijano Otero. The first was a *Relacion y detalle de la Provincia de Santa Maria la Antigua del Darien* signed by Don Andres de Ariza on the 12th of September 1778. This report and a map that accompanied it were addressed to "Al Excelentisimo Señor Don José de Galvez, del Consejo de Estado y Secretario del ReaI y supremo de Indias." The report says:

> The above mentioned mine of Espiritu Santo has been and is the greatest in the world. Its rich metal is sold by measure, not weight. Its principal vein is said to be seven yards wide. Its wealth was so great, that the cupidity and lust of foreigners were so roused that the province was plundered many times and finally reduced to its present wretched condition. According to trustworthy reports, the gold received by the King in 1708 (the royal 20 per cent) from the mines of Cana, amounted to 82 thousand castellanos (820 pounds).

Ariza used several expressions that were in vogue in Darien to describe the richness of the Espiritu Santo mine, such as that "the mine produced more gold than earth," and that the "metal was measured in it by the bushel."

The other manuscript is a report, made by order of his Majesty, upon a project formed by Don Jeronimo Sancho, Adjutant Major of the military department of Panama, to operate the Espiritu Santo mine. This report was addressed by Señor Ariza to his Excellency Don Manuel Antonio Florez, Archbishop Viceroy of Santate, and was dated June 28th, 1780. It contains an interesting description of Cana.

> Cana forms a valley four or five leagues in extent, and surrounded by high and beautiful mountains to which it owes its cool climate; this valley is longer from north to south, where numerous rivers unite to form the great Tuira.
>
> The soil of the valley is the most fertile in the province; coffee, cocoa and indigo are grown here, as well as the products of cold climes.
>
> The great abundance of gold in the mines of Cana astonishes one; but nature seems to have added to the marvel by the manner in which she has distributed it, the rivers all flowing near the deposits, and thus facilitating their operation.
>
> It is a well known fact that there is in Cana an immense quantity of gold in veins, small and great, not only in the Espiritu Santo mine, so celebrated the world over for its magnificence, but also in the deposits that originate there, and spread out from it in a continuous and almost interminable series.
>
> In less than 15 years, and solely through its abundance of gold, Cana was transformed from a desolate mountain into a large and thrifty town, the largest in the province. And in its ruins today may be seen not only ditches, bridges and remains of other works which were constructed for operating the mines, but also many furnaces, streets, &.c, that attest how numerous must have been its population.
>
> The mines of Cana were discovered in the last years of the XVIIth century, up to which time the most famous mines of the province had

> been those of Troncoso, Bagre and Tayecua, in the district of Tucuti. As soon as news was received of the abundance of gold in Espiritu Sarnto, as well as of the ease with which it could be mined, most of the miners of Tucuti and elsewhere flocked to the new field.
>
> But on account of the troubles, which began immediately and continued for several years; troubles with the English, the French and the Indians, the town was absolutely ruined, and the mines of Cana, and the rich and fertile deposits of Tucuti and its neighborhood, as far as the Gulf of San Miguel and the Chocó, were all abandoned.

Restrepo (1886) described another report signed by Ariza in 1790 in the archives of the hydrographic depository of Madrid, called a *Descripcion de la Provincia del Darien.*

Spanish Mine Description

Ariza's report of the mine exploitation was confirmed when the mine was reopened about 170 years later (Restrepo, 1886, 1888; Woakes, 1895, 1899). The Spanish mine exploited a steeply inclined orebody by means of a shaft with 4 ladders with 12 to 15 steps each, and four levels to a depth of 60 m, the deepest being the most extensive. Mine support was by pillars and timber, but gradual robbing of pillars was a problem. The bags of ore were hauled up the shaft by a wheel or winch operated by two men at the bottom of the shaft. These two men were killed when the shaft caved in. The ore was processed by panning to recover free native gold. The mine was kept free of water by five wooden treadwheels.

The mine was worked by 200 miners who were mostly free blacks, although there were also slaves. They worked in day and night shifts for six days a week. They were paid one pan of gold ore per day which they panned on Sundays. This would give between 16 and 50 castellanos of gold, or 2.4 to 7.4 Troy ounces, today worth between USD11,520 and USD35,520 (at USD4,800 per ounce).

Historical Maps

Cana is shown on numerous historical maps between 1697 and 1814 but none after 1814 except for Wyse (1877). It is named variously as "Gold Mines and River" and "Santa Maria" (1697, 1733 map), "Las Minas de Santa Cruz de Cana" (1744), "Real de Santa Maria", "Camino Real de Cana", "Cana", "Minas Altas", "Cerro del Espiritu Santo" (1747), "Cana" (1761), "Mina del Espirito Santo" and "Santa Cruz de Cana" (1765), and "Santa Cruz de Cana" (1778), "Cana" and "Espiritu Santo de Cana" (1814).

Las Minas de Santa Cruz de Cana detail of 1744 map.[15]

6. Explorers in the 19th Century

Emile Lebreton (1840-1852)

In 1840, a wealthy Parisian scholar, Mr. Emile Lebreton, became strongly interested in the gold mines of Santa Cruz de Cana (Restrepo, 1884, 1886). He organized a French company, sailed for Panama, and spent some time in Darien, where the explorer Lionel Gisborne saw him in 1852 (Gisborne, 1853). His enterprise was not successful.

Irish and English Explorers: Edward Cullen (1852) and Lionel Gisborne (1853)

Dr. Edward Cullen, an Irish doctor, traveler and a Fellow of the Royal Geographical Society, announced in a book published in 1852, with a much enlarged second edition in 1853, the discovery of a low level route across the Darien from Caledonia Bay to the Gulf of San Miguel which he claimed would be the ideal route for a ship canal (Cullen, 1852, 1853). It proved to be fraud.

His claim was supported by Lionel Gisborne, who supposedly surveyed Cullen's trail, in his book *Journal of the Expedition of Inquiry for the Junction of the Atlantic and Pacific Oceans* (Gisborne, 1853). Gisborne was the chief engineer of the Atlantic and Pacific Junction Company of London that was formed to promote the canal route. Both Cullen and Gisborne had an interest in the company and their claims were clearly promotional. Gisborne wrote that a French company had concessions on the River Grande and their plan was to open the gold mines at "Cannes". The problem was the shortage of labor (Gisborne, 1853, p. 173). He hired a small 12 ton schooner to take them from Panama to San Miguel. The owner of the schooner was Dr. Le Breton who was the agent to the Gold Mining Company that had a concession at the "Cannes" mines (Gisborne, 1853, p. 175).

Dr Wozencraft (1870s)

In the 1870s, a well-known spiritualist from the United States, Dr. O. M. Wozencraft, said he had received in New York a communication which induced him to go to Panama in search of the long lost mines of Cana (Restrepo, 1884, 1886). After having visited Darien, he published an interesting description of his adventures and misadventures in the *The Californian* literary magazine under the title of "The Wilds of the Darien" in three parts (Wozencraft, 1881). It seems that he was unable to find the lost mine, and he soon returned to California.

Lucien Napoleon Bonaparte Wyse (1876-1877)

The mine and the ruins of Santa Cruz de Cana[16] were visited in 1876-1877 by the French Navy Lieutenant Lucien Napoleon Bonaparte Wyse (1846-1917), the engineer surveying routes in the Darien for the French canal company (Wyse, 1877; Restrepo, 1884, 1886). He was the grandson of Napoleon's brother Lucien Bonaparte. "La Société Civile Internationale du Canal Interocéanique de Darien" (1876-1880) was better known as the Türr Syndicate after the lead director, the Hungarian General Istvan Türr. The other two directors were Baron Jacques due Reinach, a financier, and Lucien Wyse, brother-in-law of Türr. Wyse obtained the 'Wyse Concession' to build a canal from the Colombian government and, as a result, the company was bought by the flamboyant Ferdinand de Lesseps' Compagnie Universale in 1880 (Wyse, 1877; McCullogh, 1977, p. 59-66; de Banville, 2005).

Wyse was guided to Cana by Don Francisco Rojas and by the report of Don Andrés de Ariza. He surveyed the mine area. He observed two artificial ponds 35 m long constructed for the purpose of operating the Espiritu Santo vein and descended to the lower part of a small waterfall. He wrote:

> From this point on, the auriferous vein runs from WSW to ENE, then

> from SW to NE as far as the entrance to the mine, which was closed by the terrible accident of 1727. The vein reappears a little further off, and it would be desirable to excavate towards the SW. (Wyse, 1877).

Lt. Wyse staked the mines of Espiritu Santo and Santa Rosa, and several placers, in the name of the French Canal Company, La Société Civile Internationale du Canal Interocéanique de Darien, but did not complete the applications and they were not granted. Wyse wrote enthusiastically about the proposed Cana mine.

> The climate of the beautiful plain of Cana is truly delicious. The fertility of its soil is remarkable, as is evident by the flourishing plantations of the Indians. The abundance of water makes the cultivation of various articles a success; and everything seems to proclaim that if in the future attempts should be made to mine the great mineral wealth of the district, its very situation provides all the facilities that could be desired for that purpose. (Wyse, 1877).

Wyse and Reclus surveyed a remarkably detailed topographic relief map of the Darien (Wyse, 1877). The map shows the Plateau de Cana with Mount Pirri on the NW side, Mount Setetule on the NE side and Loma Espiritu Santo on the south. The plateau has a small lake and is traversed by the River Tuyra from the north which is joined by the River Seteganti from the SW which exits the plateau to the SE through a canyon called the Salteria de Peña Hueca. Santa Cruz de Cana is shown with '*Ruinas*', having been abandoned 150 years before. The River Cana flows east into the plateau to join the Seteganti with several very short named tributaries - Rosario, Santa Rosa, Espiritu, and Pobres – the number of named features shows how important the area was for mining. To the north the Rivers San José – with a small lake, Escucha Ruido and Lemon flow east into the Tuira.

The Wyse and Reclus map was redrawn and used by Woakes (1899) for the mine.[18]

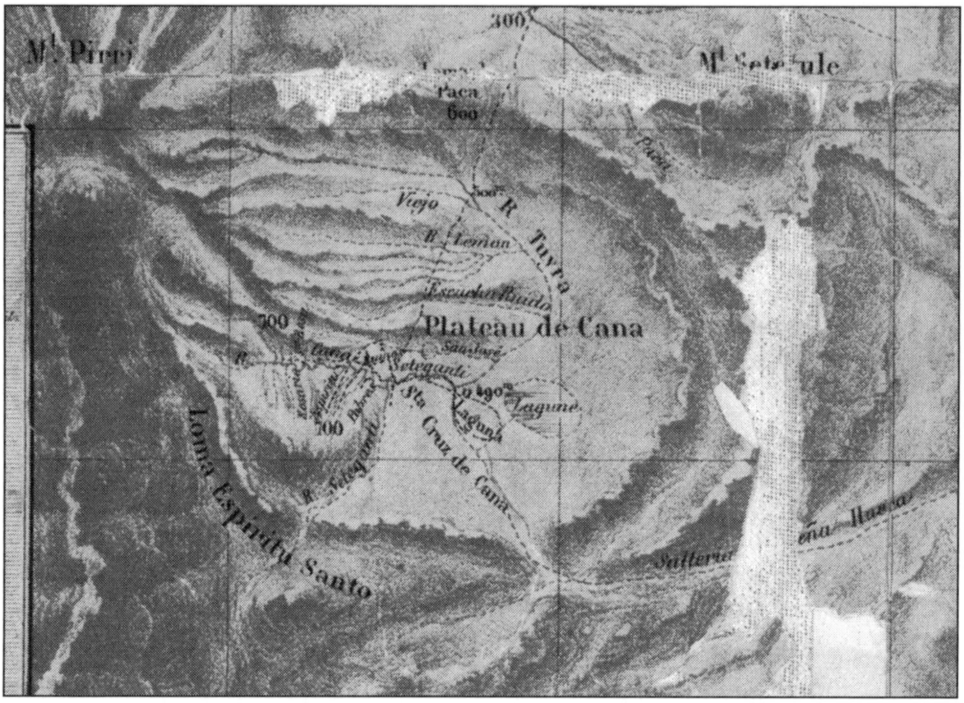

Transactions of the American Institute of Mining Engineers.—Vol. XXIX. WOAKES.

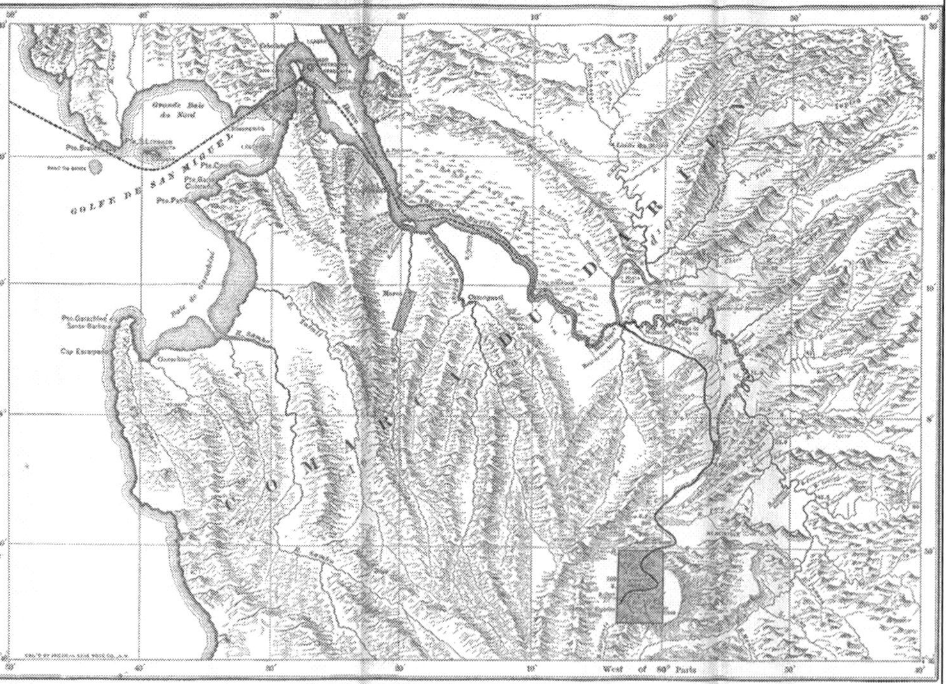

MAP OF THE DARIEN, SHOWING ROUTE TO THE MINES.

(Copied by L.D. Inglis, from the Map of Lucien N.B. Wyse.)

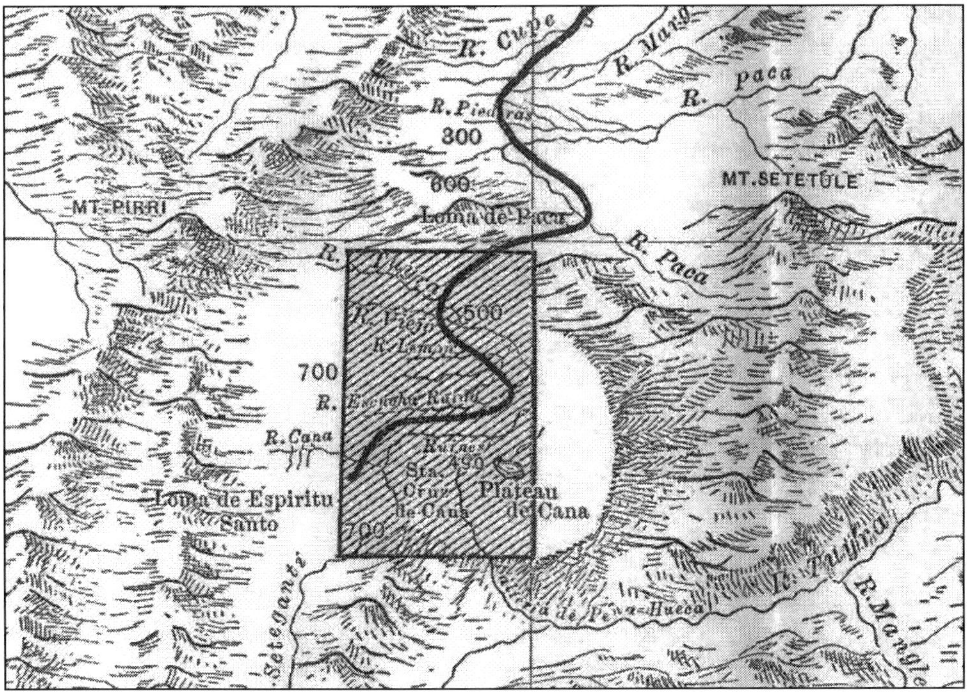

Above: Detail of the Cana mine from the map by Woakes (1899).

Opposite page top: Santa Cruz de Cana and the Cana Plateau detailed enlargement from the Wyse and Reclus map of the Darien, 1877 (Wyse, 1877) The map scale is 1:250,000. The grid lines are 5' or 5 nautical miles equals 9.25 km.[17]

Opposite page bottom: Map of the Darien by Wyse (1877); redrawn by Woakes (1899).

Armand Reclus (1876-1877)

Lt. Armand Reclus (1843-1927) was the second in command of the French canal company and wrote an important memoir (Reclus, 1881). He visited Chepigana on the River Tuira, sailing from Panama on the small steamer *Taboguilla*, but did not visit the Cana mine. He wrote that:

> The geological riches of this region are incalculable. The gold mines of Cana, although poorly exploited, were the most productive in Central America.[19]

The Gold Mines of Cana Company (1878-1880)

Mr. Newcomb, a friend of Mr. Wozencraft, and a spiritualist like him, persevered in the search for the lost gold mine, and organized a new company in the State of New York in 1878 called The Gold Mines of Cana. It had an authorized capital of $40,000 in 400 shares of $100.[20] The president was Mr. H.S. Wells and the secretary was E.S. Wells. Mr. Wells and other New York capitalists were stockholders in it, and Mr. Normand Rude, a North American, who had travelled through Chocó and part of the Isthmus, was placed in charge of the operations.

Mr. Rude transported to Cana a large number of hydraulic machines, pumps and other mining equipment for the operation of a quartz vein; then he spent the entire capital of the company, more than $50,000, in putting up a saw-mill, in opening roads and preparing the way. Just when he believed he had reached the old galleries of the mine, the saw-mill broke down and the entire project was abandoned (in the beginning of 1880), without having made any examination of the vein, nor extracting a grain of gold.

The Wells Company offered to sell its rights to another American company, which sent to Darien an engineer, Colonel J. H. Crossman. The latter soon learned that Mr. Rude had not found the Espiritu Santo vein and made an unfavorable report. Mr. Crossman, however, did enter the old gallery, which was still open; he found there some heavy timbers in a good state of preservation, and was able to penetrate to the third level. It will be remembered that it was the shaft alone which caved in in 1727, and that the miners were able to escape by the galleries. But he returned home without exploring the mine at all, because he was employed to value it as he found it, and not to discover it.

7. The Lost Gold Mine Reopened

Compañía Minera del Darien (1883)

The rediscovery of the Espiritu Santo de Cana gold mine in 1883, after being abandoned for almost two centuries, is another remarkable story. The Spanish accounts of the fabled, rich gold mine aroused the interest of Vicente Restrepo, a Colombian historian, metallurgist and government minister (Restrepo, 1884, 1886). He was carrying out research to annotate the travels of Lionel Wafer in the Darien and came across the manuscript by Lieutenant Colonel Don Andrés de Ariza, former Governor of Darien, who described the mine in detail. He obtained the accounts of pirates and other authors from England to verify Ariza. Restrepo wrote,

> though for twenty-five years, I had closely watched the progress of mining industries in the State of Antioquia, I knew that its very richest mines had never yielded the incredible quantities of metal that had been extracted from the mines of Darien.

He determined to find the mine. He ascertained from Don Francisco B. Rojas of Panama that the mines were abandoned. Restrepo requested the mine concession from the Government of Panama on behalf of the Compañía Minera del Darien owned by himself and a group of friends, with the help of Sr. Rojas. He made a short trip to Paris and made an exploration and operation agreement for the mine with Mr. M. S. Maisounabe, an old friend who had access to capital. The mines were inaccessible with no road. An expedition was set up for Sr. Rojas to take possession of the mine by chartering a boat to take Sr. Rojas and horses from Panama to Pinogana on the

River Tuira. From there he took the District Judge, his secretary and surveyor to the mine in order to take possession.

Don Francisco B. Rojas (1870-1883)

Don Francisco B. Rojas was a Panamanian miner who lived in the Darien and who knew the location of the Cana mine. He guided Wyse and The Gold Mines of Cana and was awarded shares in the latter for his services, and Restrepo enlisted his help (Restrepo, 1884, 1886). Rojas had visited the mines several times and had written a pamphlet titled *El Darien y sus terrenos auriferos*. Restrepo described him thus:

> [as having] had a profound belief, almost a blind faith, in the riches of Cana; for many years he has had an idea, which he has not yet abandoned, which was to discover and operate the Espiritu Santo mine. No one is better acquainted with the region of Cana and its environs. Heis a man of integrity, is intelligent, active and resolute, and much respected throughout Darien." (Restrepo, 1884, 1886).

Rediscovery (1883)

Gold and other minerals in the ground in Colombia and Panama belong to the state and an individual or company can be granted a concession giving the right to explore and mine these minerals with payment of a royalty, taxes and annual license fees. The final part of the legal process to be granted a mining claim is to take physical possession of the mine in the presence of the mining authority, or in this case the local judge, and verify the marker posts, no matter how remote it is. It required an expedition to take possession of the Cana mine, as described by Restrepo (1884, 1886):

> We have thus far reproduced this account just as we published it in "El Repertorio Colombiano," [No. 2, October, 1883]. It only remains to complete it. Señor Francisco Rojas, charged with taking possession of the Espiritu Santo mine, sailed from

Panama, July 6th, 1883. On the 11th, he disembarked at Santa Maria, a village situated on the left bank of the Tuira, from which place he journeyed by land to Pinogana. Here he was to take the road that leads to Cana, but found it entirely closed. It was necessary to open it through its entire length, (some 30 leagues).

Overcoming many difficulties, he at last reached the place where the "Wells American Company" had had its work, on the right bank of the Cana river. On this spot, he found the shaft that had been examined by Colonel Crossman. Near the mine was the house built by Mr. Rude, and a large building in which was the sawmill. As Señor Rojas had been informed that the seam at this place was a branch of the Espiritu Santo deposit, and not the mother vein, he proposed to seek the latter.

He approached the Cana river, and followed its course to the east; after passing it, he ascended a little hill, and from there descended to the old shaft, which measured 120 metres in length and 33 in width. **

On the borders of the shaft he found four tolerably deep wells, which it was not possible to examine on account of water dripping from their interiors. It is probable that through these openings the descent to the lower galleries of the mine was made. Going west from the old shaft, he reached the slope of the Espiritu Santo hill, and there had the satisfaction of seeing and ·measuring an enormous vertical vein, nine metres wide ***.

On the 15th of October, Don Francisco Rojas returned to Pinogana, where he was awaited by the Political Judge of the district and the other personages who were to assist at the formal taking possession of the mine.

His followers went to the plain of Cana, where they employed themselves for four days in measuring two claims in the mine, one in each of the two shafts that have been discovered. By means of picks they broke out from the vein several specimens of quartz, which contained auriforous pyrites. On their return they brought from the mines of Cana, two cannons, more than a metre in length, made in Barcelona in 1700.

Señor Rojas returned to Panama the 10th of December, after an absence of five months.

Señor Rojas doubtless had knowledge of an old manuscript "Noticia de las minas que se conocen en el Istmo de Panama," (published in the previous number of these Annals), in which the following passage occurs: "3,000 paces from Cana, is the famous royal mine ** * * which yielded an unheard of abundance of gold. A short distance from the royal mines, is a deposit called Espiritu Santo, which is a not her very rich gold mine.

** What Señor Rojas calls a shaft, is more properly a cut or small opening into the vein.

*** This vein was not touched by the former inhabitants, as they considered it very poor in gold. Specimens of its ore were assayed in Paris and only gave 15 grammes of gold and 55 of silver, to the ton.

In 1869 Señor Jose Manuel Groot published in the first volume of his *Historia Eclesiastica y Civil de Nueva Granada*, a summary of the important report upon the wealth of the mines of Cana, addressed to the Viceroy Guirior by Don Andres de Ariza.

To the honor of our country, be it said, it is a Colombian, in the service of a Colombian company, who has at last succeeded in discovering the rich Espiritu Santo mine, buried for a century and a half.

For this discovery let us give all the well deserved credit and praise to Señor Don Francisco B. Rojas.

The Darien Gold Mining Company Limited (1887-1891)

The formation of The Darien Gold Mining Company was carried out through a rather complicated series of transactions which are described here.

30 December 1882[21]: Contract of Purchase by Mr. Saturnin Maisounabe signed before Mr. Mare and his Colleague, Notaries Public of Paris. Ratified by Mr. Ramirez Castro, Notary of Bogota,

9 March 1883. Maisonnabe was the owner of four lode concessions for two lode mines Veta Madre de Espiritu Santo and Mina del Rey granted on 23 January 1884, and three placer mines called Champs de Mars, Champs Elysees and Bois de Boulogne. The titles were being completed.

20 February 1887[22]**:** An 1887 Agreement in French with English translation deals with the French holdings, properties and so on. Mr. Saturnin Maisounabe was the owner of four lode concessions for two lode mines Veta Madre de Espiritu Santo and Mina del Rey granted on 23 January 1884, and three placers called Champs de Mars, Champs Elyses and Bois de Boulogne. A Commission was formed to evaluate the mines and make a map to determine their value. Maisounabe was to get 40% of the new company. Georges Michael Braggiotti and Auguste Figge of London intended to form a new company. Signed Paris, 20 February 1887. London, 21 February 1887.

2 March 1887: In 1887 Restrepo sold the Compañía Minera del Darien to The Darien Gold Mining Company Ltd., with registered office at 2 Victoria Mansions, Westminster, England, an Anglo-French company based in 86 King Street Manchester, England, with its Memorandum of Association of 2 March 1887. The authorized capital was £200,000 in 200 shares of £1. The shareholders were mostly French. It was managed by the industrialist Richard Hammersley Heenan of Heenan and Froude Ltd., Manchester and whose English directors included Lord Charles Beresford, Colonel Edis of the Artists' Rifle Volunteers, and Arthur Coventry of Smith and Coventry Ltd. (Restrepo, 1888; Woakes, 1923).

March 1887: On 3 March 1887, the company, represented by George Dirs Mertens, made an Agreement to purchase the interest in the gold mines of Cana, Darien held by Georges Michael Braggioth, Augustus Figge and Richard Hammersley Heenan. The price was £135,000 paid in shares.[23] The mine concessions had been granted by Decree of the Governor of Panama on 23 January 1884.

25 March 1887: Indenture[24]: Amended the agreement of 20 February 1887. Richard Hammersley Heenan was granted two paid parts of eight in the company.

14 December 1887: A Special Resolution of 14 December 1887 approved new regulations based on the deal between the French and Heenan.[25]

THE COMPANIES ACTS, 1862 to 1883

COMPANY LIMITED BY SHARES.

Memorandum of Association

OF THE

DARIEN GOLD MINING COMPANY,

LIMITED.

The Memorandum of Association of the Darien Gold Mining Company, 1887.

The registered office was changed on 23 January 1888 to No 1 Chancery Lane, Pall Mall, Manchester.[26]

The company was liquidated voluntarily in 1891 starting on 14 September after the assets were transferred to a new company. By 1919 the company had been wound up but not liquidated, due to a lack of funds to pay for publication of notices.[27]

Vicente Restrepo Maya (1837-1899)[28]

Vicente Restrepo Maya was a Colombian chemist, metallurgist, businessman, writer, historian, archaeologist and government minister, who is best known for his book *Estudio Sobre las Minas de*

Vicente Restrepo Maya (Restrepo, 1952).

Oro y Plata de Colombia (*A Study of The Gold and Silver Mines of Colombia*) (Restrepo, 1884, 1886, 1888, 1937, 1952) and for rediscovering the Cana gold mine. Born in Medellin, he studied in France at the Hermanos de las Escuelas Cristianas in Passy. He then studied chemistry at the Laboratory of Pelouze and Tremy and learned how to analyze minerals and metals. On his return to Medellin in 1857, he set up the first gold and silver refinery and assay laboratory with his brother Don Pastor. In 1879 he moved to Bogota to write. He served as Minister of Foreign Relations (1884-1887, 1888-1889), Minister of the Treasury (1885, 1886, 1886-1888, 1890) and Minister of Hacienda (1888) under President Rafael Núñez, three times president of Colombia (1880-1882, 1884-1886, 1886-1894). He published the first edition of the Mines book while in office in 1884: it was a sensation, caused a gold rush, and was translated into French, English, and Italian. He ended up with 500 shares in the Darien Gold Mining Company that he founded which he disposed of on 15 July 1890.[29] As a historian he translated Lionel Wafer's Darien journal of the surgeon-pirate (1888), and wrote biographies of Jimenez de Quesada (1897), Archbishop Arias de Ugarte and others. In the field of archaeology, he studied the muisca calendar and wrote a highly regarded history of the Chibcha before the conquest (1895). He dedicated the last 8 years of his life to the Society of San Vicente de Paul. He died in Bogota in 1899 aged 62.

Saturnine Maisounabe

Monsieur Saturnine Maisounabe of Paris, described by Restrepo as "an old friend who had access to capital", raised the start-up capital. Restrepo had studied in France until the age of twenty in 1857 and must have met him then, a quarter century before. Nothing is known about him. Restrepo made a trip to Paris to visit him in 1882. The French canal construction was in full swing since 1881 so the country was popular with French investors, before the financial problems, fraud scandal and bankruptcy of 1889. A Contract of Purchase by Mr. Saturnine Maisounabe was signed before Mr. Mare and his Colleague, Notaries Public of Paris on 30 December 1882. It was ratified by Mr. Ramirez Castro, Notary of Bogota, on 9 March 1883. Maisounabe became the owner of applications for four lode concessions for two lode mines called Veta Madre de Espiritu Santo and Mina del Rey that were later granted on 23 January 1884, and three placers called Champs de Mars, Champs Elysees and Bois de Boulogne. But it took him more than four years to raise money for the company, and the money came from London rather than Paris. The company had a lot of French shareholders and French directors controlled the board from 1905.

Richard Hammersley Heenan (1847-1920)

The Darien Gold Mining Company Ltd. was based in Manchester, England, and was managed by the industrialist Richard Hammersley Heenan of Heenan and Froude Ltd. The English company directors included Lord Charles Beresford, Colonel Edis of the Artists' Rifle Volunteers, and Arthur Coventry of Smith and Coventry Ltd. (Restrepo, 1888; Woakes, 1923).

Richard Hammersley Heenan was born at Birr, Ireland in 1847 and educated at Parsonstown, County Ofally.[30] He worked in India from 1864-1880 on the East Indian Railway, then in the Public Works Department and later for Bawhipore State. On his return to England, he set up Heenan & Froude in Manchester in 1881, an engineering company which built the Blackpool Tower in 1884. He married Ada Drummond. He was a member of Manchester City Council from 1902, and later of Colwyn Bay District Council, and was appointed as a magistrate for Denbighshire, Wales in 1906. He retired in 1918 and died in Knightsbridge, London in 1920 aged 73. He was survived by his wife Ada and five children. The company lasted until 1986.

Ernest Woakes, FIMM (1864-1947)[31]

Ernest Woakes was an English mine engineer for the Darien Gold Mining Company for seven years from 1892-1899 who rediscovered the Espiritu Santo mine and made the mine profitable. He published two technical papers about the mine while working there (Woakes, 1895, 1899) and later he wrote about his reminiscences in Woakes (1923).

Ernest Russell Woakes (1864-1947) was educated at University College School, London, and received his professional training at the Royal School of Mines from 1882 to 1885, where he graduated with the A.R.S.M. in mining. In 1886 he was appointed assayer and assistant engineer to Tolima Mining Co., Ltd., Colombia, and subsequently became assistant manager. He left in 1891 and in the following year took up the position of manager to Darien Gold Mining Co., Ltd., Panama, where he remained for seven years.

From 1899 to 1903 he held the post of resident consulting engineer in British Columbia to a number of properties and then went to Spain for Messrs. John Taylor & Sons to assume the managership of Linares Lead Mining Co., Ltd., Alamillos Mining Co., and Fortuna Mining Co., in 1904 receiving a Diploma from the Spanish Government for his work. He managed other lead and copper mines in Spain until 1910.

In that year he became consulting mining engineer in London to Messrs. John Taylor & Sons, a position which he retained until his retirement in 1942. In 1911 he went to Namaqualand for a short period to manage the mines of Cape Copper Co. In subsequent years he visited and reported on gold mines practically all over the world, and on lead, copper, wolfram, vanadium and molybdenum properties in many countries. In 1918 he was in Norway on behalf of the British Government in charge of work for the restriction of the enemy supply of molybdenum. In 1923 he was in India reporting on the copper mines which were acquired by the Indian Copper Corporation Ltd., to which company he remained a consulting engineer until 1932. From 1921 to 1932 he was a director and latterly chairman of Balaghat Gold Mines, Ltd., one of the mines on the Kolar Gold Field of South India. He was president of the Royal School of Mines (Old Students') Association from 1919 to 1920. He was elected to Associateship of the institution in 1893 and transferred to Membership in the following year, and he served the Institution as a Member of Council for the 28 years from 1915 to 1943. He died on 29th November 1947, at Stevenage, Hertfordshire, at the age of 83, after a long illness.

Mr. Thomas Pryor wrote: With characteristic self-effacement, the article by E.R. Woakes in the Mining Magazine of November 1923, on the Darien gold mine, Panama, makes little mention of the part played by himself in that romantic and successful venture. Yet the whole credit for its success is due to him and to him alone: it was his careful research into the old records which showed that the previous unsuccessful prospecting in the area had never found the richest of the old Spanish workings, and it was his refusal to he daunted by formidable difficulties which led to the re-discovery and opening of the old mine with such conspicuous financial success. These qualities of personal modesty, professional integrity, technical competence and determination were displayed throughout his long career.

The Darien Gold Mining Company Limited (1891-1907)

A new company was formed with Articles of Association on 22 September 1891 and was incorporated on 23 September 1891 with the same name, The Darien Gold Mining Company, Limited.[32] An agreement was made between the old and new companies to restructure and increase the share capital on 25 September 1891. The authorized capital was £100,000 in 50,000 A shares of £1 and 50,000 B shares of £1.

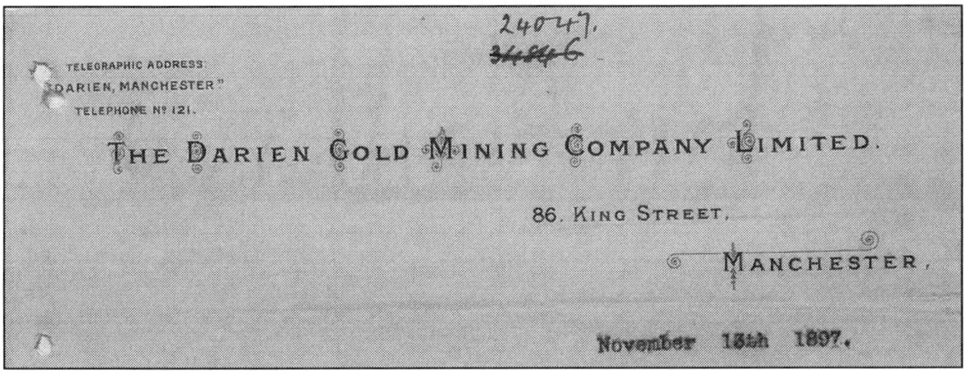
24047.

TELEGRAPHIC ADDRESS
"DARIEN, MANCHESTER"
TELEPHONE Nº 121.

THE DARIEN GOLD MINING COMPANY LIMITED.

86, KING STREET,
MANCHESTER,

November 18th 1897.

The Darien Gold Mining Company letterhead, 1897.

It increased the share capital on 19 April 1893 by £25,000 in 25,000 C shares of £1. It issued a debenture on 16 January 1903 for £50,000 in 2,500 debentures of £20 paying 10% to 1915. It issued a trust deed for £50,000 in 10% debentures, part of an authorized issue of 75,000, on 16 January 1905 with a Prospectus.[45]

William E. Proctor of California was the Mine Manager from 1903, and Gerald W. Lachszyrma was the Chief Engineer (Oller, 1933).

The Rediscovery of Espiritu Santo Mine (1893-1896)

The first three years of the company are described in newspapers reports of the annual general meetings.

In 1892 Hammersley Heenan presided over the meeting of the Darien Gold Mining Co. in Manchester (The Times, 19 January 1892).

In 1893 Hammersley Heenan presided over a meeting of the Darien Gold Mining Co in Manchester at which the shareholders were told of the state of the mines; he reminded them that the stories of Spanish gold extracted from the mine had never been disproved. Further capital was needed for machinery. The chairman suggested the speculation would be a good one if carried through. (*The Glasgow Herald*, 31 January 1893).

In 1894 Hammersley Heenan presided over a meeting of the Darien Gold Mining Co in Manchester; amongst other directors present was Lord Charles Beresford. The company had been searching for gold for 5 years and they had now struck a lode 40 feet wide, "the widest in the world". The company had sufficient cash to last until crushing of ore started in the following year. (*Leeds Mercury*, 6 November 1894).

The company initially mined the El Rey Mine (or North Mine) for two years until the end of 1892 with a 20-stamp mill, erected in 1889 (Chibas, 1898) but it was low grade with poor results. The original capital was exhausted as well as £30,000 raised in the restructuring. The company brought in the English mine engineer Ernest Woakes as mine manager from 1892-1899. He rediscovered the Espiritu Santo Mine or South Mine. An exploration shaft was sunk to 80 feet depth and a crosscut was driven under the open cut workings. In January 1893 it cut old workings and the pumps and shaft "drowned with hardly a minute's warning" (Woakes, 1896). Additional capital was put up. A drainage tunnel 1,080 feet long was started in July 1893 to intersect the mine at 90 foot depth. In the next dry season, the shaft was de-watered with a Cameron sinking pump. The shaft was sunk 10 feet more and a drift was started which met the adit on 3 July 1894, and the water drained out to the Cana River. Levels and cross-cuts were driven into the lode which was 80 feet wide from north to south and formed by conglomerate. It started to process ore from pillars and debris in May 1895 using a 10-stamp mill . It installed pumps and winding machinery in 1895 with a new 270 foot deep engine shaft called Heenan "after our sanguine president", with a double winding engine and two skips. The 100 foot crosscut below the adit cut the lode on the north side in December 1895. After driving south for 30 ft. through low-grade conglomerate ore, rich rock was met together with a tremendous stream of water. The water suddenly diminished and another 4 or 5 ft. of driving took them into old workings. Here they found five wooden treadwheels used by the Spaniards to dewater the mine, preserved under water for 170 years (Woakes, 1895, 1899, 1923). Woakes (1896) described the spectacular sight.

> These [old workings] were for the most part open, and could be explored to a certain extent, and I believe such a sight as met our astonished eyes has seldom, if ever, been seen by men now alive.
>
> The old workings consisted chiefly of enormous shafts reaching upwards into darkness. They were most elaborately and systematically timbered. In two of these hanging 40 ft. above our heads were wooden treadwheels, two more were in position at our 100-ft. level, and in so perfect a state of preservation that one of them was turned a bit by a venturesome miner before it broke. These wheels are carefully made of cedar wood, are 12 ft. diameter, having a wooden axle with iron gudgeons and bands. Mounted on the axle is a drum about 4 ft. diame-

> ter, round which, no doubt, an endless chain and buckets of some sort were carried to wind the water from one stage to the other. No sign of these chains are to be found, only some pieces of tanned leather and laces. The remains of a similar wheel had been found at the adit level. Launders and bateas, hewn from the solid, are plentiful, also ladders, made by cutting notches at regular intervals in an 8-in. diameter pole. These are placed in pairs side by side, the steps "breaking joint," so to speak, and being placed at a convenient angle in the shafts, are easy to ascend. No doubt all the ore was carried up there in bateas by slaves. The poor beggars must have had a hard time of it on the treadwheels. We are pumping 80 gallons a minute to the adit. They, no doubt, had to raise the same amount a great deal higher. (Woakes, 1896).

The shaft was sunk another 100 foot in 1896, the woodwork of the second 10-stamps was renewed to be able to crush 1,200 to 1,500 tons a month, and a 6 mile long ditch was cut from the Seteganti River to give a head of 330 feet above the main adit to drive a hydraulic motor for the pumps (Woakes, 1896).

Eduardo Chibas gave a detailed description of the plant in 1898 (Chibas, 1898). The ore was extracted by the Heenan Engine Shaft using compressed air for hoisting, drilling and pumping. It was generated by a Schram-duplex air-compressor driven by a water-wheel constructed on the Pelton principle with a head of 90 feet of water. The ore was taken in cars hauled by mules over a ½ mile long tramway and bridge. The ore was dumped into a 100 ton ore bins and by four automatic feeders to four Fraser and Chalmers batteries of 5 stamps weighing 85 lbs. each. The ore was free milling with recovery of about 70% by amalgamation. The 'sulphurets' (sulfides) were kept for treatment by a cyanide plant erected in 1898. The stamp mill was operated by a Pelton wheel with 96 foot head of water that was increased to 200 feet in 1898. The Pelton wheels generated electricity also.

In April 1896 the mill crushed 493 tons and recovered 1,190.88 oz gold giving a recovered head grade of 2.43 opt (Woakes, 1896). Production in 1896-1897 was 10,528 ounces sold for £44,522 at an average grade of 1 to 1.5 opt (Chibas, 1898). The mine was profitable for several years and the company paid an average annual dividend of 23.3% between 1892 and 1903 (Rippy, 1953; Fischer, 1975). Annual gold production from 1896 to 1907 was between 10,528 ounces and a record 36,671 ounces (1903-1904).[34]

Albert Fraissignes, the French shareholders' lawyer, made a visit to the mine in 1901 and wrote a book about it (Fraissignes, 1902; quoted in Araúz, 2005b).

FIG. 1.

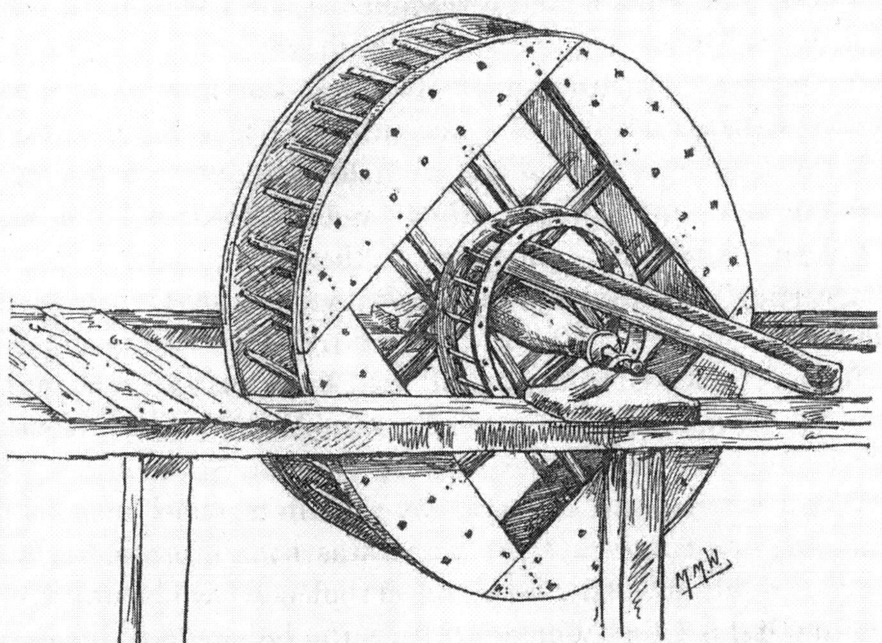

Old Spanish Tread-Wheel, Found in Espiritu Santo Mine, Cana.

170-year old wooden tread wheel found in e Espiritu Santo de Cana mine (Woakes, 1899).

Fig. 2.

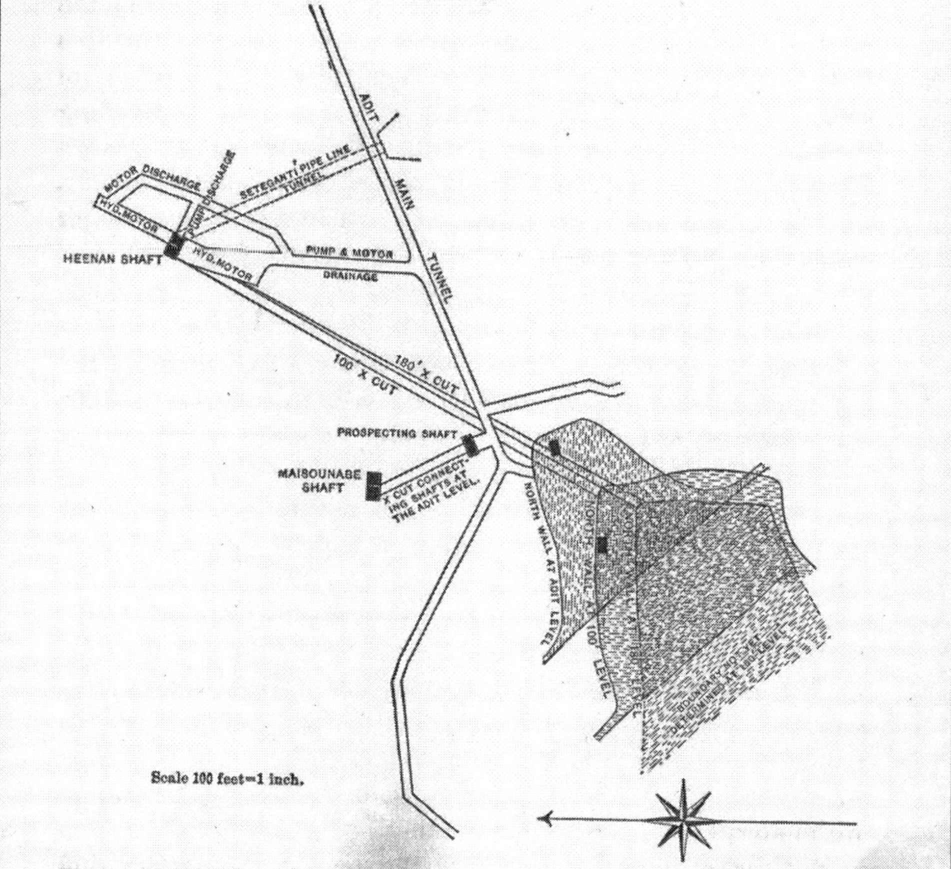

Plan of Espiritu Santo Mine, Cana, Re-worked by the Darien Mining Co., Ltd.

Plan of the Espiritu Santo de Cana mine (Woakes, 1899)..

Fig. 3.

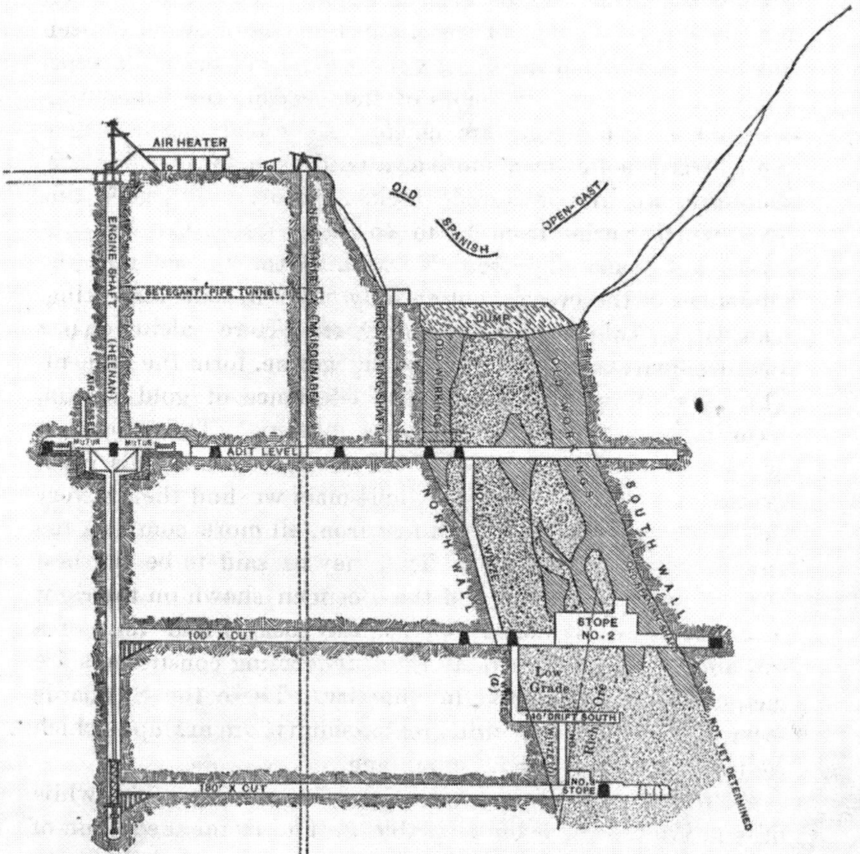

Scale 100 feet=1 inch.

Section of the Espiritu Santo Mine, Cana, Re-worked by the Darien Mining Co., Ltd.

Cross section of the Espiritu Santo de Cana mine (Woakes, 1899).

José Oller Navarro (1882-1984)

José Oller Navarro (1882-1984) was the Administrator of the mine in the early twentieth century. He described himself as Office

Jose Oller Navarro (Germinal Sarasqueta).

Assistant (Oficinista Auxiliar) (Oller, 1975). His collection of photos forms the basis of this book. He wrote an article about *La Industria Minera de Panama* (The Mining Industry of Panama) in 1933, most of which was about the Cana mine (Oller, 1933), with an update in 1975 as a short book with an expanded text and photos of the mine (Oller, 1975). He married Griselda Pinel, the daughter of Próspero Pinel, the brother of Pablo Pinel, the legal representative of the company in Panama (of whom more below).

Bert Harrison

Another collection of photos came from Petern Tristam, the great nephew of Bert Harrison in Australia.[35] Bert Harrison was an accountant from Manchester who worked at Cana for 3 years from about 1900 and then went to Australia. He was married to Peter Tristam's great aunt who was his dad's mum's twin sister but they had no children. The photos were salvaged by Peter's father, scattered on the floor after his house had been stripped following his death; it was raided by 'friends' then thieves and all of his artifacts were stolen including spears, a huge snake skin, minerals, etc. He was in his 80s and living on his own in a small town called Milton about 200 km south of Sydney.

Peter's mother remembers him talking about travelling in mule trains and being terribly sick with various fevers but it was certainly the highlight of his life. Peter writes, "I remember him as a very old man when I was a kid and his stories of jungle, monkeys, minerals and Indians certainly had an effect on my future passions!"

Darien Monetary Tokens

The first Darien Gold Mining Company issued its own coinage or tokens for the use of employees at the company store.[36] These were large brass octagonal coins minted by George Yorke Iliffe of Birmingham in England in denominations of 10, 20, 25 and 50 cents and one dollar. Three series were made. The first is dated 1894, the second series is undated, and the third series was struck in 1903-1905. The third series has the token maker's mark "G.Y. ILIFFE BIRM." The tokens are octagonal except for the one dollar which is square and the third series 10 cents which is hexagonal. The obverse has a miner with shovel in the mountains with the inscription 'The Darien Gold Mining Company Limited', and the reverse has the denomination and the inscription: '*Para simplificar las cuentas nada mas*' ('To simplify the accounts, nothing else').

Darien Gold Mining Company Ltd tokens
(Above: courtesy of David S. Plowman.
Below: courtesy of Nicolás Liakópulos Falcón).

The Thousand Day War (1899-1902)

The mine's steamer *Darien* was taken over by the rebels during the Thousand Day War (1899-1902) and was used in the rebel naval fleet at Panama (Chibas, 1898; Woakes, 1923). A Navy list gives the name as *Panama*, ex-*Darien*,[37] that was a gunboat bought by the government on 17 June 1901, was seized by the rebels on 4 November 1901 (*Darien*), was returned to the government after the war on 21 November 1902 and was transferred (*Darien*) to the Panamanian Navy on 6 November 1903. The Panamanian army and navy were abolished by law on 20 November 1904 and the navy ships were sold, but the fate of the *Darien* is not known.

The Independence of Panama (1903)

Panama declared its independence from Colombia on 3 November 1903 and the mine became part of the new republic. The frontier was only 5 km from the mine. The American employees left the mine fearing an invasion by the Colombian army to recover Panama (Woakes, 1923). Oller (1975) wrote that the administrative and technical staff left the mine due to rumors of an invasion by a Colombian army unit camped near the border.

The US sent to warships to patrol the River Tuira to prevent an invasion from Colombia. Sheridan (1926) wrote:

> All of the quartz and placer gold properties can be reached from Panama City by gasoline or oil-burning boats, to the Gulf of San Miguel, and then up the Tuyra River to Real de Santa Maria and Yaviza. The U. S. Gunboat *Yorktown* made the trip as outlined above, and anchored off Yaviza for some time, landing 250 marines, during a rebellion against Colombia in 1903.

The warship was actually the monitor USS *Wyoming* (1902-1937) accompanied by the cruiser USS *Boston* (1887-1946) which "would steam up the Tuira River and land marines in support of the pro-United States Panamanian insurrection." The USS *Boston* and

Above: USS *Wyoming* circa 1919 after being renamed *Cheyenne*.[40]

Below: USS *Boston* in 1891.[41]

USS *Wyoming* must be the largest ships to have sailed up the River Tuira and would have been an imposing sight.

The USS *Wyoming* was an Arkansas-class monitor also known as USS *Monitor No. 10* or USS *BM No. 10*. She was built by the Union Iron Works, San Francisco, in 1902. Her dimensions were 255 ft (77.7 m) length by 50 ft (15 m) beam by 12ft 6 in (3.8 m) draft, displacement 3,225 long tons, and she was armed with 2 x 12 inch guns, 4 x 4 inch guns and 3 x 6-pounder 57 mm (2.2 inch) guns. She was renamed *Cheyenne* in 1909 and was withdrawn from active service in 1937.

The USS *Boston* (1887-1946, renamed *Dispatch*, 1940) was a protected cruiser, built by the Delaware River Iron Ship Building and Engine Works, Chester, Pennsylvania. Her dimensions were 283 ft (86.3 m) length by 42 ft (12.8 m) beam by 17 ft (5.2 m) draft, displacement 3,189 long tons, powered by steam and sail. She was armed with 2 x 8 inch guns, 6 x 6 inch guns, 2 x 6 pounder 57 mm 2.24 inch guns, 2 x 3 pounder revolving cannon, 2 x 1 pounder revolving cannon, and 2 x .45 caliber Gatling guns.

The US Naval History and Heritage Command describes the USS *Boston*'s movements:

> However, when the Colombian province of Panama declared its independence on 3 November 1903, *Boston* received orders to the new republic's western coast, ostensibly to protect American interests should fighting ensue. She arrived off Panama on 7 November, and the remaining ships of the Pacific Squadron joined her three days later. The presence of United States naval vessels on both coasts of Panama had the effects of insuring the success of the insurgents and of allowing the United States to acquire from them a treaty for the construction and control of a transisthmian canal.
>
> *Boston* remained in Panamanian waters until 2 March 1904 at which time she sailed for Callao, Peru, in company with *New York*, *Bennington*, and *Concord*. The warships arrived at Callao on 9 March and remained there until the 25th. *Boston* returned to Panama on

1 April and stayed there until 26 May at which time she put to sea bound for San Francisco. The warship arrived in the California port on 12 June. She remained there until 1 August when she resumed normal operations along the California coast. By mid-November, *Boston* found herself operating along the west coast of Panama once more. That assignment occupied her time until 8 February 1905 when she headed back to San Francisco.[42]

The USS *Wyoming* was new having been built in San Francisco in 1902 and fitted out at Mare Island. She ran trials and exercises in California in 1902-1903 and sailed south. Robert J. Cressman picks up the story:[43]

> ... reaching Acapulco, Mexico, on 31 October. She subsequently shifted further south, to Colombia, where a civil war and a rebellion in the Panamanian territory instigated by the United States government, took part in the independence of the Panamanian territory from Colombia, to later build the Panama Canal. The monitor accordingly arrived in Panamanian waters on 13 November and sailed up the Tuira River in company with the protected cruiser *Boston*, with a company of Marines under Lieutenant S. A. M. Patterson, USMC, and Lieutenant C.B. Taylor, USMC, embarked, to land at Yariza and observe the movements of Colombian troops.
>
> The presence of American armed might there and elsewhere ultimately aided in independence for the Panamanians. During that time, *Wyoming* anchored at the Bay of San Miguel, Panama, on 15 December. The following day, a boat with 11 Marines embarked for the port of La Palma, Panama, under sail. While *Boston* departed the scene on 17 December, *Wyoming* shifted to La Palma on the following day. There, Lieutenant Patterson, USMC, with a detachment of 25 marines, commandeered the steamer *Tuira* and took her upriver. While the Marines were gone, a party of evacuated American nationals came out to the monitor in her gig.
>
> Meanwhile, Patterson's Marines had joined the ship's landing force at the village of Real to keep an eye on American interests there.

Back at La Palma, *Wyoming* continued to take on board American nationals fleeing from the troubled land and kept up a steady stream of supplies to her landing party of bluejackets and Marines at Real. Ultimately, when the need for them had passed, the landing party returned to the ship on Christmas Eve. *Wyoming* remained in Panamanian waters into the spring of 1904 keeping a figurative eye on local conditions before she departed Panama Bay on 19 April, bound for Acapulco, Mexico.

The present border was defined by the Colombian Law of 9 June 1855 when the State of Panama (1855-1886) was created within the Republic of New Granada (1831-1858). It was renamed the Granadine Confederation in 1858 and the United States of Colombia in 1863 and was centralized again as the Republic of Colombia in 1886, and the State of Panama was reduced to a department. Following the separation of Panama on 3 November 1903, Colombia and the United States signed a treaty on 6 April 1914 which recognized the 1855 Colombia–Panama border. Colombia and Panama signed a border treaty on 20 August 1924, and the border was demarcated on the ground by 14 boundary-marker pillars in 1938.[44]

The mine became subject to the laws of Panama rather than Colombia. Production was good until the mine flooded in 1907 (Oller, 1975).

William E. Proctor (1903-1907)

The production was satisfactory with the best year being the financial year ending in February 1904. This was thanks to the manager, William E. Proctor, a Californian miner from Sutter's Creek where the California gold rush started. Production was from the Espiritu Santo mine. He made a new tunnel to level 9 at 900 feet depth, the deepest level, drilled an inclined hole from surface to level 4 at 400 feet depth for safety to communicate with the surface, and extended the tunnels. He was ably helped by Juan Manelia and the head mine engineer, Gerald W. Lachszyrma (Oller, 1975: 6).

Block Caving (1904)

Block caving was introduced at the Espiritu Santo mine in 1904 by a young mining engineer from San Francisco called Norris Nathan English, who was acting manager (Chase, 1907). This is one of the earliest uses of block caving known. The upper levels of the mine were high-grade running 2.35 ounces per ton gold and required a lot of timbering in square sets which was expensive and inefficient. The cut-off grade was 1 ounce per ton and anything less was left behind unmined or as fill. The grade was significantly lower below the 5th level so a lower cost mining technique was required. The orebody on the 6th level was triangular in shape with sides of 150 by 120 by 80 feet (45 x 36 x 24 m). The 6th and 7th levels were caved by undermining them 14 feet (4.3 m) below by four raises, drifts cut on all three sides and six crosscuts. The roof of the blocks was blasted until the old workings were encountered on the level above. The pillars were then mined and the whole floor of the level was covered in broken or which was then extracted. The pit created on the surface with filled with waste rock.

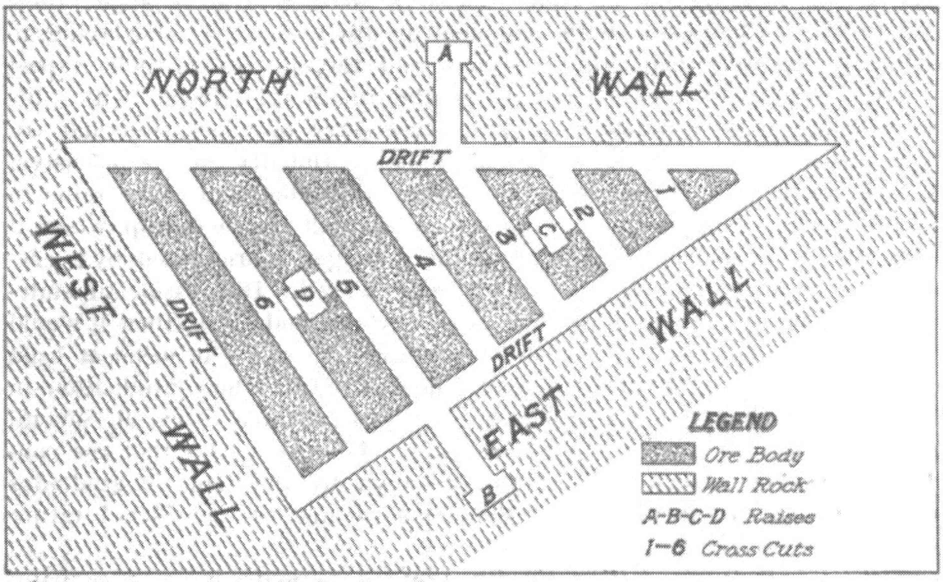

Plan of the caving system of the 6th level of the Espiritu Santo mine in 1904 (Chase, 1907).

A Mine Tour

José Oller, the administrator at the mine, wrote a description of an underground visit to the Espiritu Santo mine from between 1903 to 1906:

> Perhaps our readers will be interested in the story of a trip to the bottom of a mine, if that mine is known to some of them even by name; and perhaps this story will be curious if it is a gold mine on Panamanian soil, for whom vernacular things have some value.
>
> We were very amused by the question that engineer Lachsyrma asked us, when we were putting on the high rubber boots, the rubberized cape and the waxed hat with which we were preparing, following his example, to go down to the mine. Have you made your will, Mr. Oller? At this half-serious, half-jovial intimation, our spirits were somewhat overwhelmed, and we answered him: I have no assets and, furthermore, since I see that you go down to the mine frequently and I do not know that you have made a will, it seems useless to me that I do so now... And we set off. We climbed into one of the large buckets where the materials and tools for work and the mineral are carried from top to bottom and vice versa, and on the edges of which the workers hold on, holding on with both hands isolated from the rigid lateral cables that join at a certain height at a central point from which the bucket hangs.
>
> Once the signal is given by means of the special thin cable called the telegraph cable, which makes a bell ring, the powerful hoist, like a mythological monster lying on the surface of the ground near the entrance or shaft, with an inexhaustible belly from which tentacles of endless extension emerge, rapidly unwinds the nine hundred feet of steel cable that it has wound around its cylindrical body.
>
> The progress of the bucket that takes us through the different levels or floors of the mine is dizzying. The sensation it produces is indescribable. Atomic fragments of earthy and moist substance hit us in the face, and as we move at this almost terrifying speed towards the last level, we see distant little lights shining from the tunnels, which emerge from their caverns in one direction or another around us.

At a touch of the signal telegraph, our perpendicular car stops on the fifth floor. There we get off the bucket and walk among the timbers that support the square, this is the impression that one receives, of more than fifty feet in length around. Admirable in its structure, this square is formed by pillars of hard and resistant local wood, such as guayacán and others of equal qualities, which are fitted one on top of the other up to a certain height, and from one side to the other, forming a curious series of columns supported one on the other. A wonderful ensemble, which demonstrates the work of man who is rooted in science to apply it to architecture even in the depths of the earth! From there we make an excursion along a tunnel, to the end of which we arrive, where we see a team of miners who drive the tips of their picks into the stones recently loosened by the action of the dynamite discharges during recess hours, or else they handle the shovel to throw the smaller mineral into the wagons that will then have to be pushed by hand to the junction of the entrance, where again the mineral will have to be thrown into the large buckets for its extraction to the surface.

Upon reaching the bottom of the shaft, the air is rarefied, with a slight smell of oil. As an experiment, a torch that shines serenely is carried to a point where it threatens to go out due to lack of oxygen. The work of the large air compressors from the surface sends, like colossal lungs, the air towards the depth of the mine, making the miracle of life in the bowels of the earth. With a pickaxe that we carry by hand, a worker takes us to a place where the gold vein or vein is more visible, and there we admire the gold in its natural and raw state. On the sinuous and cracked wall of the blueish rock, the quartz shows its treasure. Innumerable filaments or compact pyrites reveal the rich vein of precious metal that man eagerly pursues, risking his life. From there, a specimen is given to us that later ends up in the furnace crucibles of the assay laboratory.

We ride again on the bucket that will take us to the ninth and last level. The force of vertigo takes hold of us, and we finally reach the bottom of the mine. Our clothes are dampened by the water seeping through the rock of the rich isthmian soil. Mysterious water that when it filters through the earth masses of the caves forms the wonderful structures that no artist could ever imitate. Our rubber boots splash, and in a click we walk along the dark tunnel, which threatens to flood, and by the light of the lamp that we carry by hand, high up, we continue along a tunnel where breathing becomes very difficult because of the rarefied air. Here the double miracle of air compressors and high-power centrifugal pumps becomes apparent, so that without these machines of industrial science, modern mining could not take place to exploit mining properties at thousands of feet of depth, as happens in the Transvaal, for example, where the depth of some gold mines has reached colossal proportions.

While in this last level, we suddenly felt the sensation of fluttering: something strange, as if the wings of a bat were passing over our heads. We were not given an explanation of the phenomenon, of this unpleasant sensation at such depth, until once outside, going through the serious range of the impressions received, Mr. Lachszyrma explained to us that it was due to the discharges of drill holes, the explosion of dynamite on other levels and in tunnels far from where we were. The thin air, due to the distance and the interposition of a mass of earth and rocks, did not allow the sound to be perceived, no matter how tremendous it was! The journey back from the heart of the earth to the surface is like that of an express train, without stops, where the transition from semi-darkness to full light, from rarefied, heavy air, artificial air if we could call it that, to the open air that the lungs eagerly welcome, is a sensation that makes us enjoy life with joy, that makes us return to the beauty of forms, of colour, when we can contemplate again the greenery of the tropical mountains of our isthmus, and their undulations, evoking the gracefulness of women, or their majesty before the serene sky in broad daylight... The will would have been useless, Mr. Lachszyrma, we said to the phlegmatic engineer, once we were out of the danger that he had made us suppose before going down to the mine. And so it is in modern mines, where every precaution is taken, except against the unforeseen, an unavoidable factor that, like a tragic question, always hangs over the heads of mortals. (Oller, 1975, p. 5-7).

8. The French Mine (1905 - 1912)

French Coup (1905)

French shareholders owned most of the shares by 1905. The French board replaced the American mine engineer managers with French engineers who made radical changes in mining methods, and increased the stipends considerably (Oller, 1975, p. 5).

The company issued a trust deed of £50,000 in 10% debentures, part of an authorized issue of £75,000, on 16 January 1905 with a prospectus.[37] The 1905 prospectus showed there were now 3 English directors and 6 French ones. The English directors were Hammersley Heenan, Chairman; Arthur Coventry; and Thomas Tatham. The French directors were A. Lodin de Lepinay, Vice Chairman; A. Chaumier, L. Maisounabe, Vct. Ch. de Leusse, L. Mellier, L. P.G. Delbruck. The registered office in 1898 was at 86 King Street, Manchester.

Hammersley Heenan resigned as a director on 13 March 1905. Thomas Tatham ceased to be a director on 24 March 1905 because he did not hold the necessary qualification of shares. New directors were installed on 29 May 1905 who were mostly French. The two English directors were Arthur Coventry, Reigate, Chairman and William T. Stark, Salisbury. The seven French directors were Arthur Lodin de Lepinay, Arsene Chaumier, Louis L.A. Maisounabe, Vicomte Charles de Leusse, Lucien Mellier, Luc P.G. Delbruck, and Emile Brerault, all of Paris.[46] Louis L.A. Maisounabe is presumed to be the son of M. Saturnino Maisounabe who was a friend of Restrepo and raised capital for the first company in 1882-1887.

The Darien Gold Mining Company Limited Act 1905 (5 Edw. 7) authorized capital of £250,000 in 250,000 shares of £1 each and eliminated the previous A, B and C shares.[47] The capital was increased by £125,000 on 22 September 1905. A mortgage of £24,000 was obtained on 15 January 1906. At an extraordinary general meeting (EGM) on 13 August 1906 and 7 September 1906, the board approved the voluntary liquidation of the company and to form a new company. The liquidator gave his consent to form a new company with the same name on 18 January 1907. The company was finally dissolved by notice in the *London Gazette* on 18 January 1921.[48]

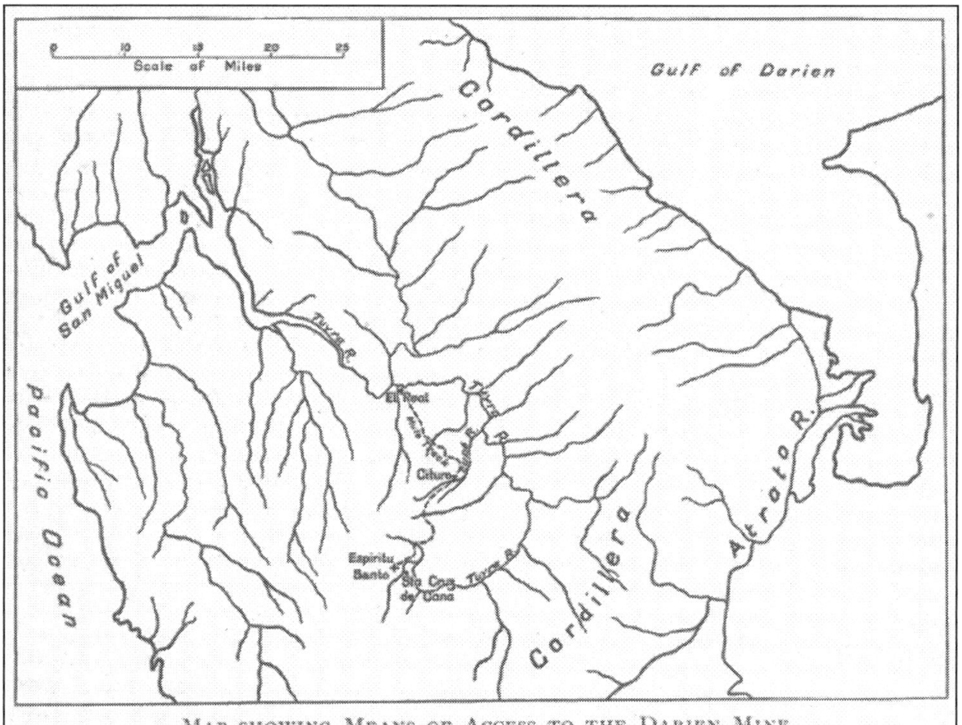

MAP SHOWING MEANS OF ACCESS TO THE DARIEN MINE.

Map showing access to the Cana mine (Woakes, 1923).

The Darien Gold Mining Company Limited (1907-1912)

Production dropped in 1907 due to flooding of the mine. The directors declared voluntary liquidation in 1907 and raised new capital in France to revive the mine with a new French-run company.

However, the mine stopped production in 1912 and the company declared bankruptcy in 1913 in the difficult pre-war economic situation in France.

A new company was set up in England on 18 January 1907, the third one with the same name, The Darien Gold Mining Company, Limited.[49] It had an authorized share capital of £350,000 in 200,000 ordinary shares of £1 each and 150,000 preference shares of £1 each. The new company was now controlled by the French shareholders and was run by French directors and managers, with the sole exception of the president Arthur Coventry of Reigate. The French directors were Emile Brerault, Arsene Chaumier, Lucien Mellier, Henri Prosper Rameau, Marius Rousset, and Jean Vincent Emmanuel Thiollier. The Prospectus was in French and English.

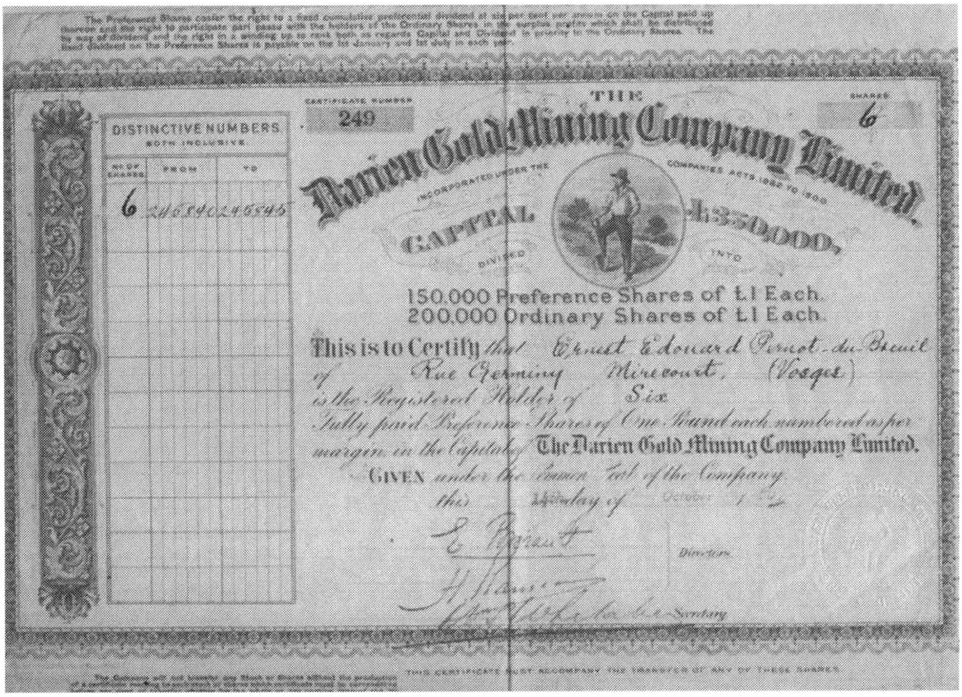

DISTINCTIVE NUMBERS
249
THE
Darien Gold Mining Company Limited.
CAPITAL £350,000.
150,000 Preference Shares of £1 Each.
200,000 Ordinary Shares of £1 Each.
This is to Certify
The Darien Gold Mining Company Limited.
GIVEN

Stock certificate of the new Darien Gold Mining Company, 1907.
(Vicente Pascual Collection).

The mine suffered from a lack of capital from 1906. The company spent about £500,000 in four years on mine development and equipment equal to the total sales. The result was disastrous. In the year ended February 1907 production fell to 6,000 ounces for £20,000. The company went into voluntary liquidation (Oller, 1975: 8-9). On 12 October 1908 it issued 1,000 debentures of £100 paying 6%, and on 31 October 1910 it issued another 2,000 debentures maturing in 31 December 1918.

The first task of the restructured company in 1907 was to build the light railway. The company restarted mining operations in 1908 with the mining engineer M. Degoutin as manager. Production was from Cerro de San José and the Santa Barbara mine. A new plant was installed with a 'Clero' mill in 1908 which was replaced by a 'Comambault' mill in 1911. The plant was changed several times and used 'Chileno' mills with vertical wheels and 'Huntington' mills but with mixed results (Oller, 1975: 9). The restructured company carried out mining at Cana from 1908-1912. Mining was suspended in 1912 and new capital could not be raised. In late 1912 a mission formed by M.E. Brérault and M. Michel arrived in Darien. They focused their attention on the placer gold mines of River La Marea in the district of Chepigana which belonged to the company; the results were good, but the company was unable to raise capital in pre-war Europe. In 1913 the America engineers Eugene Lefevre and James Anderson visited the mine but with no results. The company went into voluntary liquidation on 10 March 1913.[59] It was fully wound up on 20 May 1919.

The production from 1895-1910 was given as 230,443 ounces (7,168 kg) by Tower (1963) at grades of 45-70 g/t, and for 1900-1907 it was recorded as 142,236 ounces (4,424 kg) by Oller (1933).

A lawsuit of Thiollier vs Darien was filed at the Chancery of Lane, Manchester on 30 November 1911 and was not heard until 3 July 1917.[51] Jean Thiollier, a former director of the company, and Claude Baillourd represented themselves and other holders of mortgage debentures. The Court ordered the receivers to pay the remaining

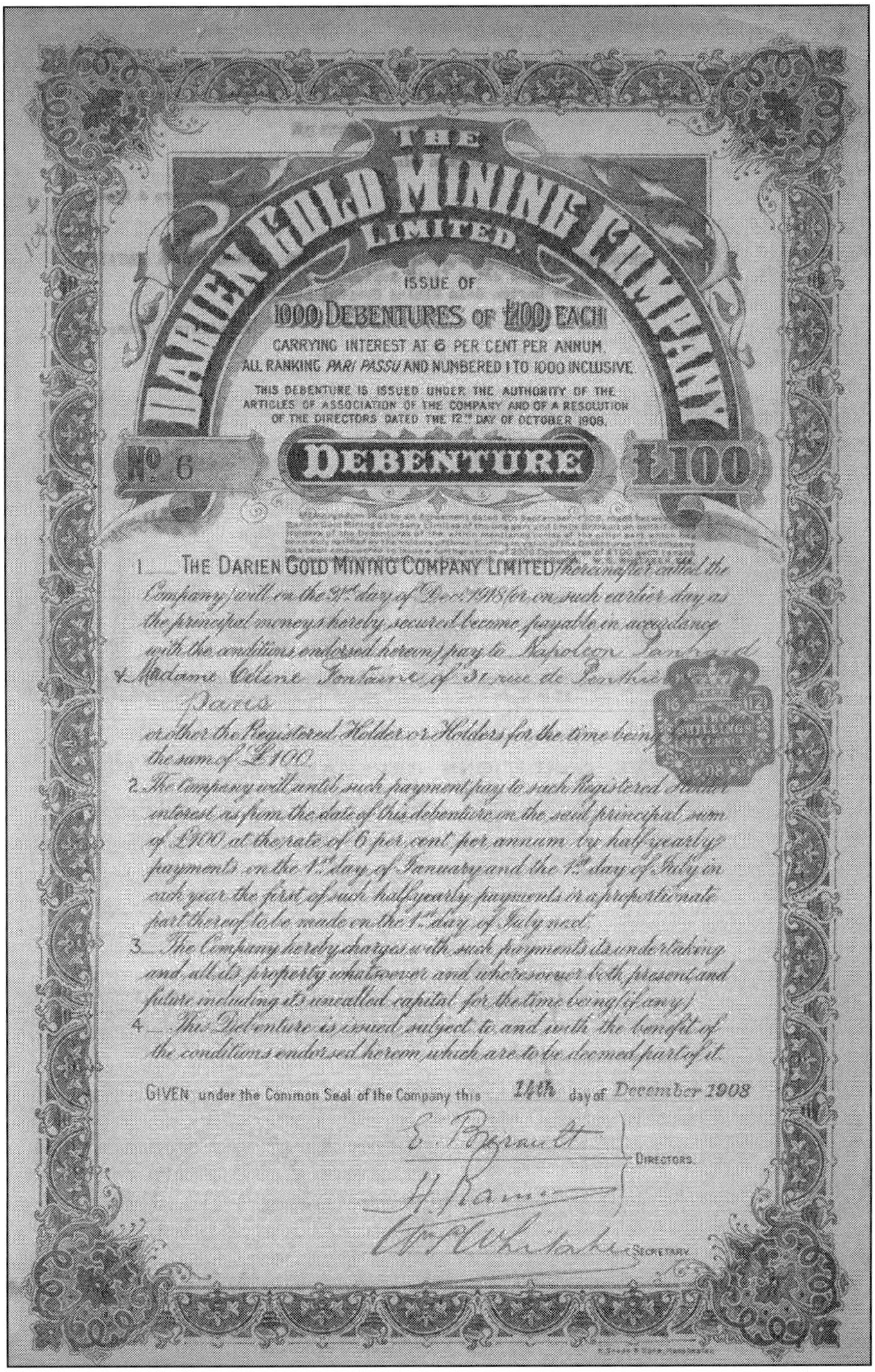

THE
DARIEN GOLD MINING COMPANY
LIMITED

ISSUE OF
1000 DEBENTURES OF £100 EACH
CARRYING INTEREST AT 6 PER CENT PER ANNUM.
ALL RANKING *PARI PASSU* AND NUMBERED 1 TO 1000 INCLUSIVE.
THIS DEBENTURE IS ISSUED UNDER THE AUTHORITY OF THE ARTICLES OF ASSOCIATION OF THE COMPANY AND OF A RESOLUTION OF THE DIRECTORS DATED THE 12TH DAY OF OCTOBER 1908.

No 6 DEBENTURE £100

1. THE DARIEN GOLD MINING COMPANY LIMITED (hereinafter called the Company) will on the 31st day of Decr 1918 (or on such earlier day as the principal moneys hereby secured become payable in accordance with the conditions endorsed hereon) pay to Napoleon Lambard & Madame Celine Fontaine of 31 rue de Penthie… Paris or other the Registered Holder or Holders for the time being the sum of £100.
2. The Company will until such payment pay to such Registered Holder interest as from the date of this debenture on the said principal sum of £100 at the rate of 6 per cent per annum by half yearly payments on the 1st day of January and the 1st day of July in each year the first of such half yearly payments or a proportionate part thereof to be made on the 1st day of July next.
3. The Company hereby charges with such payments its undertaking and all its property whatsoever and wheresoever both present and future including its uncalled capital for the time being (if any).
4. This Debenture is issued subject to and with the benefit of the conditions endorsed hereon which are to be deemed part of it.

GIVEN under the Common Seal of the Company this 14th day of December 1908

DIRECTORS.

SECRETARY.

Debenture £100, 1911 (Courtesy Mario Boone, wwww.ttcdeinze.be).

cheques to the addresses supplied, then the debenture could be cancelled and the receivers discharged. The court was solemnly informed that the cheques had indeed been sent to France on 31 May 1916 but that many were returned with the stamp or a note stating, "addresses in invaded country" or "addressees were either dead or could not be found."

Captain J.F. Sheridan of New York wrote of the Espiritu Santo mine in 1926 (Sheridan, 1926):

> In spite of much interruption in milling due to caving of shafts and stopes and to other causes, the bullion produced from 1899 to 1903 inclusive amounted to nearly $2,500,000, and the depth reached was about 500 ft.
>
> Because of the caving of the stopes a large amount of ore still remained above the fifth level; so much so that over $500,000 was afterwards recovered by milling this mixture of ore and waste rock. The orebody is a prismatic chimney with a horizontal area of about 100 x 100 ft. and inclined about 18 deg. from the vertical. It is formed at the intersection of two fractures. The history of the mine from 1903 to 1911, when the mine was permanently closed, due to caving-in of the working shaft is simply a story of the incompetent management. Little or no prospecting in depth was done; other than sinking an inclined shaft 800 ft. to cut the ore at 1,500 ft. No attempt was made to crosscut the ore at the 800 level the tunnel or drift reaching the footwall only. The pump broke down, the mine was flooded to the 300 level, and for some unknown reason has not been unwatered, although the total seepage did not exceed fifty cubic feet per minute.

Placer Gold Mining and Disseminated Gold

Placer gold deposits at San José were mined by the Spaniards and by the French-controlled Darien Gold Mining Company (Oller, 1975, 17-18 citing Degoutin, 1913). The River Cana flows east through Cana town to San José 1 km to the east where it discharges in a plain

3-4 km wide and 8-10 km long, called the Cana Plain or Cana Plateau, which has deposits of alluvial gold. There were important Spanish mine workings at San José. The French carried out exploration and alluvial mining. Oller's photographs show placer gold mining using monitors.

The French company discovered disseminated gold in andesite at Cerro Santa Barbara near Mina del Norte and at San José. This was described in 1913 by M. Degoutin, an engineer, President and Director of the company since 1907, in a pamphlet published by the company called "Auriferous Andesite Mines" (Degoutin, 1913; quoted by Oller, 1933, 1975: 10). Three tunnels were dug at Santa Barbara and two others at San José showed they were mined to 15-30 m depth, and gold continued deeper with grades of 5-10 francs per ton, with some assays of 30 francs per ton or more. Degoutin compared the San José and Santa Barbara andesite-hosted gold mines with the Goldfields, Nevada mine newly discovered in 1903. The description is of disseminated gold in porphyry.

Rubber Plantations

Chibas (1898) observed the abundance of wild rubber trees at Cana on his visit in 1896. The rubber was exploited by the Indians. "The rubber industry was also very profitable, at one time, but the natives' inexcusable practice of cutting down the trees, instead of tapping them to take out the rubber, has greatly diminished the output," he wrote.

Mahogany and vegetable ivory were also exploited.

The Darien Gold Mining Company followed Chibas' recommendation and laid out rubber plantations (*caucheras*) in the flood plain of the River Seteganti, adjacent to Cana, in 1898-1900. These covered 1,500 hectares with one million trees "planted in perfectly straight lines at equal distances," although that number appears to be over-estimated. There was also a rubber plantation of 1,482 ha at Aruza (Mendez, 2004, p. 59-62). The River Aruza is the tributary north of the River Cupé; the mule trail to the mine crossed the River Aruza midway between El Real and Cituro on the River Cupé, and the light railway from Boca de Cupé passed through the plantation.

According to Mendez, the trees were '*caucho hebea*' which is *Hevea brasiliensis*, the Pará rubber tree native to the Amazon,[52] and not the native species, *Castilla elastica* or the Panama rubber tree, which grows from Mexico to Ecuador.[53] Seeds of *Hevea brasiliensis* were taken from Brazil to the Royal Botanical Gardens at Kew, London in 1875 and in 1876-77 were sent to the British colonies of Ceylon, Malaya and Singapore, but the first plantations were not developed in Malaya until 1900.[54] Being British, the Darien Gold Mining Company could have had access to seeds from Kew Gardens in London. The Darien plantation was set up in about 1898-1901 at the same time as the first plantations in Malaya. Rubber plantations were set up in Bocas del Toro at about the same time but it is not known which species of rubber tree was used, although there is evidence that it was *Hevea brasiliensis*.[55]

According to the Prospectus of the 1907 company,[56] the rubber plantations were a valuable asset. The prospectus referred to a report by a Mr. Drouot of 10 April 1906 which indicated that there were then about 250,000 trees up to 5 years old of 3 to 18 cm diameter and 1.5 to 6.0 m height, and that since then the number and size of trees had increased. Each tree of 8 years gave a profit of 3s to 4s per year. The rubber had been valued at 4/6d per pound. According to José Oller, at the time of the restructuring of the company, it had 18,000 hectares under cultivation at Aruza with 200,000 rubber trees of 5 years (Oller, 1975, p. 9).

The 5 to 8 year age of the trees in 1906 indicates that the plantation dates from about 1898 to 1901, and they are not referred to in any company archives before 1907. This is the same time as the first plantation was established in Malaya.

In 1912, when the new company went bankrupt, the company assets included 16,000 hectares of land under cultivation with 260,000 rubber trees of the species *Castilla elastica*, *Hevea* and *fi-*

cus, in plantations at Aruza, Seteganti and River Limon (Oller, 1975, p. 11). The inventory made by Mr. Degoutin in 1913 included plantations at Setegantí, River Limon and Aruza with 260,000 rubber trees of *Castilla elastica* and *Hevea brasiliensis* (Oller, 1975, p 19). The manager of the Aruza Rubber Plantation was Monsieur L. Drouot, and the manager of the tobacco plantation was Monsieur Ferdinand Didider (Oller, 1975, p. 21).

The rubber plantations were abandoned after the mine closed in 1913: parts of them died, parts degenerated, and the remainder were exploited clandestinely by Panamanian and Colombian campesinos during World War Two when there was high demand for rubber after the Malayan plantations were occupied by the Japanese. The rubber was exploited by cutting down the trees to tap them on the ground "in the easiest, most irresponsible and stupid way," thereby killing off what trees were left (Mendez, 2004, p. 57-59). The US government's Rubber Development Corporation organized rubber collection from wild trees from Mexico to Panama by hundreds of collectors during the war managed by Aaron King who was given leave of absence from the United Fruit Company.[57]

R. Williams, Botanist (1908)

The botanist Robert S. Williams (1859-1945) of the New York Botanical Garden visited the Cana mine in 1908 when it was still in operation (Williams, 1908; Barnhart, 1945; Heckadon-Moreno, 2006). The results were published in his monographs on mosses of Panama. He described the situation of the mine at 2,000 feet elevation at the foot of the Espiritu Santo mountains, which have peaks up to 7,000 feet. He worked in the field collecting from 17 to 27 April 1908.

E. Goldman, Zoologist (1912)

A zoologist from the Bureau of Biological Survey, US Department of Agriculture, Edward Alphonso Goldman, used the mine as a field camp in 1912 and published descriptions of it with photos in his field report (Goldman, 1912) and in the monograph *Mammals of Panama* (Goldman, 1920, p. 10-15). He was participating in the Biological Survey of the Panama Canal Zone by the Smithsonian Institution from 1910 to 1912 to document the natural history of the Canal Zone prior to construction of the Panama Canal. He arranged the visit with Mr. Pablo Pinel, the Panama agent (note Pablo and not Prospero Pinel). He travelled to the mine on the steamer *Cana* on the evening of 21 February. By early morning they were at the mouth of San Miguel Bay. The day was spent steaming up the bay and the estuary of the "Rio Tuyra" (Tuira), with short stops at La Palma and Chepigana, "native villages, built mainly of palm-thatched houses picturesquely grouped on the southern shore." They had to anchor for an hour in the mouth of the Chuqunaque River at nightfall to wait for the tide and reached Marraganti at 9.30 pm where they were welcomed by the agent, Mr. Pedro Campagnani. This was about 1.5 miles above the town of Real de Santa Maria, and near the limit of steam navigation. On 23 February they continued up the River Tuyra by dugout canoe or piragua for about 30 miles to Boca de Cupé, arriving after dark. From here there was a 30 mile tramway to Cana. The first stage to Mount Kitchener, in the foothills, was by a short train drawn by a gasoline motor. For the next 12 miles to Paca the track wound up the mountain in tight curves which "were slowly and laboriously traversed by push car, a platform placed on trucks and pushed by men from behind. In descending, the cars, allowed to run by gravity, were rather insecurely controlled by coils of rope wound on the axles." The final six miles from Paca was by a mule-drawn car. The journey from Panama took four days.

He stayed at the camp from 24 February to 11 April 1912, courtesy of the French manager, M. Masse, and directors M. Michel and M. Degoutin, who provided him with "comfortable quarters, transportation and other facilities ...." He used jungle camps around the mine and on Cerro Pirre until 22 June. He departed on the steamer *Cana* at 10 am. She made stops at Chepigana and La Palma, and "shortly before dark the little steamer began to rock unsteadily in the confused currents of San Miguel Bay, and the receding shoreline was suddenly blotted out by a torrential downpour of rain" (Goldman 1920, p. 14). They reached Panama at 11 am the next day, the return journey having taken just 25 hours.

9. Gold Production

Spanish Period (1675-1727)

Between hyperbole and poor to absent record-keeping, it is challenging to work out the historical gold production of old mines. Cana is no exception.

William Dampier called the mines "the richest discovered to date in America" (Dampier, 1697).

Wafer (1698) / Joyce (1934) related that a Spanish prisoner told him that the mines of Santa Maria yielded from 18,000 to 20,000 pounds (8,165-9,072 kg or 262,510-291,671 oz) of gold per year. He added: "But though this yield may have been greater or less than I have said, it is certain that an incredible quantity of gold is annually extracted."

Armand Reclus wrote: "The geological riches of the Darien are incalculable. The gold mines of Cana, despite the bad way they were exploited, were the most productive in America." (Reclus, 1881).

The high gold grade, estimated to have been 0.13% or 1,330 g/t (Restrepo, 1888), made the mine by far the biggest producer in the Americas at the time. Pockets of gold were said to contain 120-600 lbs. (55.2-276.0 kg or 775-8,874 ounces) of gold. In 1708, it was reported that the King's royal fifth tax was 82,000 castellanos (377.0 kg or 12,130 ounces) indicating total production for that year of about 1,866 kg (60,000 ounces); while other sources say the tax was as high as 100,000 castellanos (460.0 kg or 14,790 ounces) giving total production of 2,300 kg (74,000 ounces). Some sources gave annual production of more than 1 million pesos of gold (about 1,648 kg or 52,000 ounces) (Restrepo, 1886, 1888). It is estimated by Restrepo that the mine produced more than 30 million pesos of gold in the Spanish period which is about 48,600 kg (1,562,000 ounces) (Restrepo, 1886, 1888).

English Period (1887-1912)

Annual gold production for the second period of mining from 1887 to 1912 (25 years) is recorded in company reports and is compiled in Table 4. The production was given by the value in pounds, shillings and pence Sterling rather than Troy ounces because the gold price was fixed until 1971. Recorded production from 1896 to 1907 was 141,160.63 ounces for 9 out of 25 years, with an average of 16,685 oz per year, a minimum of 10,528 ounces and maximum of 36,671 ounces. These were the best years. Other years were either mine development or low grade ore with no profit but with some production. If production is assumed to have been zero in the development years and averaged 5,000 ounces per year for the other 11 years with no data, this adds 55,000 ounces for a total of 196,160 ounces.

The following estimates are given by other sources.

Sheridan (1926):

- Reports production in US$. Converted using US$19.75 per ounce (1792-1834) and US$20.67 per ounce (1834-1934).
- Spanish period (Don Andres de Ariza): $30,000,000 (1,518,987 oz).
- In one year, the tax was $284,000 meaning the total was $873,000 (no - tax 20% so total $1,420,000).
- 1899 first large bullion shipment $550,000 (26,609 oz) at a cost of $250,000.
- 1899-1903: $2,500,000 (120,948 oz).
- Later another $500,000 (24,190 oz) was recovered from above the 5th level.
- Total Period 2 of 171,747 oz.

Oller (1933):

1900-1907: 142,236 ounces (4,424 kg).

Tower (1963):

1895-1910: 235,612 ounces (x kg) at grades of 45-70 g/t.

Year ending February	Gold (£ s d)	Gold (Troy ounces)	Gold (£, Oller)	Notes (Mine, Source)
1888		0*		Development
1889		0*		El Rey mine
1890		0*		El Rey mine
1891		0*		El Rey mine
1892		0*		El Rey mine
1893		5,000*		Development
1894		5,000*		Development
1895		5,000*		Development
1896		5,000*		Development
1897		10,528		Espiritu Santo mine. Chibas, 1898.
1898		5,000*		Espiritu Santo mine
1899		5,000*		Espiritu Santo mine
1900	91,671 - 3 -1	23,550.71	91,671	Espiritu Santo mine. P-1905, 07
1901	43,835 -7 -5	11,261.50	43,835	Espiritu Santo mine. P-1905, 07
1902	41,031 – 1 – 6	10,542.39	43,031	Espiritu Santo mine. P-1905, 07
1903	66,970 – 4 – 1	17,205.03	66,970	Espiritu Santo mine. P-1905, 07
1904	154,418 – 8 – 2	36,670.76	154,418	Espiritu Santo mine. P-1905, 07
1905	52,163 – 10 - 3	13,401.03	52,164	Espiritu Santo mine. P-1907
1906	50,069 – 13 - 9	12,863.12	50,070	Espiritu Santo mine. P-1907
1907		5,138.09	20,000	Espiritu Santo mine, flooded
1908		5,000*		Santa Barbara, San Jose
1909		5,000*		Santa Barbara, San Jose
1910		5,000*		Santa Barbara, San Jose
1911		5,000*		Santa Barbara, San Jose
1912		5,000*		Santa Barbara, San Jose
Sub-total recorded	500,159 – 8 - 3	141,160.63	520,159	

Table 4 Gold production from the Cana mine, 1887 to 1912.
Sources: The Darien Gold mining Company Limited Prospectus 1905 (P-1905) and 1907 (P-1907),[58] Chibas (1898), Oller (1975). Totals in Pounds Sterling was converted to Troy ounces using the official UK gold price of £3-17-10 (£3.8925) per ounce (1717-1914). Assumed production* of 0 oz for Years 1 to 5 and 5,000 oz per year for other years with no data.

Years	Ore (tons)	Gold (oz)	Au grade (oz/ ton)	Au recovery (%)
1895-1901	49,208	64,952	1.32	79
1901-1902	11,999	24,074	2.00	86
1902-1903	43,779	60,707	1.388	88
1903-1906		80,710		
1910	19,075	5,169	0.271	75

Table 5. Darien Gold Mining Co. Ltd production (Tower, 1963).

Nelson (1995, 2006) gives estimated production in 1665-1727 of 1 Moz Au from 0.25 Mt of ore, and from 1887-1907 another 1 Moz Au from 1 Mt ore for a total of 2 Moz Au.

Summary

The best estimate of the gold production from the Cana mine is as follows:

- **Spanish Period (1675-1727):** estimated 1,562,000 ounces Au at 1,330 g/t Au.
- **English Period (1887-1912):** estimates between 196,160 and 235,612 ounces Au at 45-70 g/t Au.
- **Total:** 1,758,160 to 1,797,612 ounces Au (1.76 to 1.80 million ounces, average 1.78 Moz).

10. Arias Ownership (1915 - 1927)

Don Pablo Pinel (1869-1951), Legal Representative and Liquidator

Don Pablo Pinel (1869-1951), a Panamanian businessman, was the legal representative of the Darien Gold Mining Company in Panama for more than two decades from before 1893 until the closure in 1913. He replaced Don Luis Gaibrois, a prominent member of Panama's financial community (Oller, 1975). The Darien Gold Mining Company declared bankruptcy for the last time in 1912. The company went into voluntary liquidation on 10 March 1913.[51] It was fully wound up on 20 May 1919. Don Pablo Pinel was appointed liquidator in 1914 (Oller, 1933). He tried unsuccessfully to form a local Mining Syndicate with John N. Popham and Joshua Piza, mine owners in Costa Rica.

Pablo Pinel was engaged in boat building on Taboga and at Peña Prieta, Panama City; pearl diving in the Pearl Islands; and coastal shipping from Panama to Pedregal, Chiriqui with a fleet of passenger and cargo steamers from 1901 operating as Pinel Brothers (Hermanos Pinel) with Prospero Pinel from 1864-1926, the National Navigation Company from 1907, and later the Compañia de Navegacion Nacional (Sarasqueta Oller, 2010).

Don Tomás Arias (1856-1932), Founding Father of Panama

The assets of the Darien Gold Mining Company were bought by Don Tomás Arias on 30 April 1915 (Mendez, 2004). Don Tomás Arias (1856-1932) was one of the founding fathers of the new Republic of Panama, having been one of the members of the first Provisional Government Junta of 1903-1904.

The 1915 sale document was made before the Notary Primera of Panama, Don Antonio Facio, and was signed by Tomas Arias as the buyer and William Thomas Pullen and Frederick Nathaniel Kimball as representatives of the Association of Creditors of The Darien Gold Mining Company Ltd ("Sindicos del Concurso de Acreedores de la Darien Gold Mining Company Limited"), and Percival Heiler. The witnesses were Jose Antonio Rivera and Jose Oller. The price was B/.25,000 or £7,338 (Mendez, 2004, p. 61). The inventory is given by Oller (1933) and Mendez (2004, p. 59-62, 413). Mendez (2004) states that the mines were excluded from the sale (p. 59-62) but later (p. 413) lists 72 ½ mines (i.e. mine claims) in the sale inventory in the District of Pinogana, Corregimientos of Pinogana, Yape, Boca de Cupé including those formerly in the Corregimiento of Cana (52 mines) and District of Chepigana, Corregimientos of Camoganti, Tucuti and Chepigana (15 ½ mines). Oller (1933) says that the mines were included. Arias did not carry out mining, although Oller (1933) states that he worked the mine for 15 years; Mendez (2004) confirms that there was no production. Arias carried out ranching and reputedly kept the commissary well stocked. The Antillean, Colombian and Indigenous miners stayed on in the hope that the mine would reopen and in the meantime grew tobacco, hunted and fished using dynamite. They gradually left and by 1924 the place was deserted.

11. The Panama Corporation Limited (1926 - 1948)

Company History

The Panama Corporation, Limited was registered in London on 8 April 1926,[60] explored and mined gold in Panama from 1926-1948 and explored the Cana deposit in particular from 1927-1932. The company's first prospectus of 15 April 1926 had an authorized capital of £2 million in 2 million shares of £1.00 each. A 1930 prospectus authorized 250,000 shares more of £1.00 each. The prospectus of 22 April 1931 raised the authorized capital to £4.5 million with the issue of £2.5 million in 8% debentures.

The company was managed by Duncan Elliott Alves, Chairman, who was also the Chairman of Central America Exploration (Consolidated) Ltd. The board of directors was of high profile and comprised:[61] 1) Henry Ludwig Mond (1898-1949), later 2nd Baron Melchett, Director of Mond Nickel Company Ltd (some sources give his father as the director Lord Melchett, 1st Baron Melchett, Sir Alfred Mond (1868-1930), Chairman of Imperial Chemical Industries); 2) Andrew Percy Bennett, CMG (1866-1943), former British Consul-General in Zurich (1918-1919), British Minister to Panama and Costa Rica (1919-1923) and British Minister to Venezuela (1924-1927);[62] 3) Walter Broadbridge, A.M.I.C.E., M.I.M.M., Director, Minerals Separation Ltd and Director, Rhodesia Congo Border Concessions, Ltd; 4) Percy St. Clair Matthey, Chairman, Johnson, Matthey & Co. Ltd.; and 5) Herbert Langham Reed, Chairman, Kerr, Stuart & Co. Ltd.

Mr. John Jackson Calderwood, M.I.M.M. visited Panama on behalf of Duncan Alves in 1924 to investigate the Veraguas area. Central America Exploration (Consolidated) Ltd., Alves' private company, was granted large exploration concessions in Veraguas (1,150 square miles or 2,980 km^2 from 1924), Darien (3,400 square miles or 8,806 km^2 from 1925), Chiriqui (4,000 to 5,000 square miles or 10,360 to 12,950 km^2 in 1929), Rodriguez, on the west side of the Canal Zone (1,000 square miles or 2,590 km^2 in 1930), Ocú, and Bastimiento (Bocas del Toro) for a total of more than 27,326 km^2. The first four contracts were vended into Panama Corporation Ltd in 1926. It also had an option to purchase the mining properties of El Remance from Messrs. Graham (James Noble Graham and Jean Rose Lovett Graham of Glasgow). Hugh Marriott, a past president of the Institute of Mining & Metallurgy, became involved with the company in 1929. (Panama Corporation Ltd Prospectus, 1926;[63] Sheridan, 1926; Riddell, 1927; Letcher, 1930; Low, 1931; Norton, 1930; Oller, 1933).

The Cana Concessions Syndicate, managed by D.E. Alves, H.L. Reed and H.F. Marriott, made a contract with the Corporation which surrendered its rights on 24 March 1931 (Prospectus 1931).

The exploration concessions were really reconnaissance or prospecting permits and not mining titles. They gave Alves the "exclusive right for ten years to explore, denounce, and exploit mines" and he would receive perpetual title to each mine discovered up to a limit of 2,500 acres each (10 km^2). The concessions included all minerals except salt, petroleum, coal, and mineral waters. The enterprise was declared to be a public utility. Each contract was to be free from taxes for ten years and perpetually exempt from import duties on its machinery for mining purposes and from export duties on its products. The royalty was two per cent of the gross sales of gold. The company was permitted to develop water power, and install telephone and telegraph lines and radio stations, aqueducts, power lines, roads and tramways (Norton, 1930).

Move to Canada (1932)

The company moved its domicile from London to Montreal (Quebec) in 1932, then the financial capital of Canada (long before Toronto became the financial capital during the Quebec separatist movement of the 1960s-1970s), and was restructured as the Panama Corporation (Canada) Ltd. (Hull, 1940). The company was registered on 9 March 1932 with 4.5 million shares authorized. The English company was placed into voluntary liquidation on 28 January 1932 and was wound up on 22 February 1933.[64] The company left behind Britain gripped in the middle of the Great Depression (1929-1932), although Canada was badly affected also. Britain reintroduced the Gold Standard in 1925 which would have affected a gold mining company but withdrew it in 1931 so that wasn't the reason to leave. The reason is believed to be that the corporate tax was much more favorable in Canada, and a foreign business resident in Canada was not liable for Canadian or British income tax.

War Casualty: Gold Mines Closure (1942)

The Panama Corporation mined gold at Los Hatillos-Mina Blanca, Remance, Peñon, San Pedrito, Santa Rosa, Alto de la Mina (all Veraguas Province), El Mineral (placer at Concepcion, Veraguas), Capira, La Florida (Cacao) (Panama Province, today Panama West), Sabalo River alluvial deposit (Darien) and others in Panama (Prospectus, 1931; Redwood, 2020). It was forced to stop gold mining after 8 October 1942 when the US federal government's War Production Board issued Limitation Order L-208 which ordered the immediate closure of "nonessential" gold mines in the US in order to free up resources "that could have been better used to provide war material, especially copper" (Craig & Rimstidt, 1998). This order was applied in some allied countries such as Panama. With its revenue suddenly cut, the company was no longer viable. The Panama Corporation (Canada) Ltd. was wound up a few years later in 1948.[65] This marked the end of gold mining in Panama for four decades until 1990 (Redwood, 2020).

Suspicious Americans

Some Americans were highly suspicious of the motives of the Panama Corporation because of its near-monopoly on mining in what they considered to be their fiefdom and wondered whether the

Left: Richard Oglesby Marsh, an American engineer, explorer, diplomat, and ethnologist.[66]

Below: Panama Corporation (Canada) Limited, 312 shares of capital stock, 31 May 1938.[67]

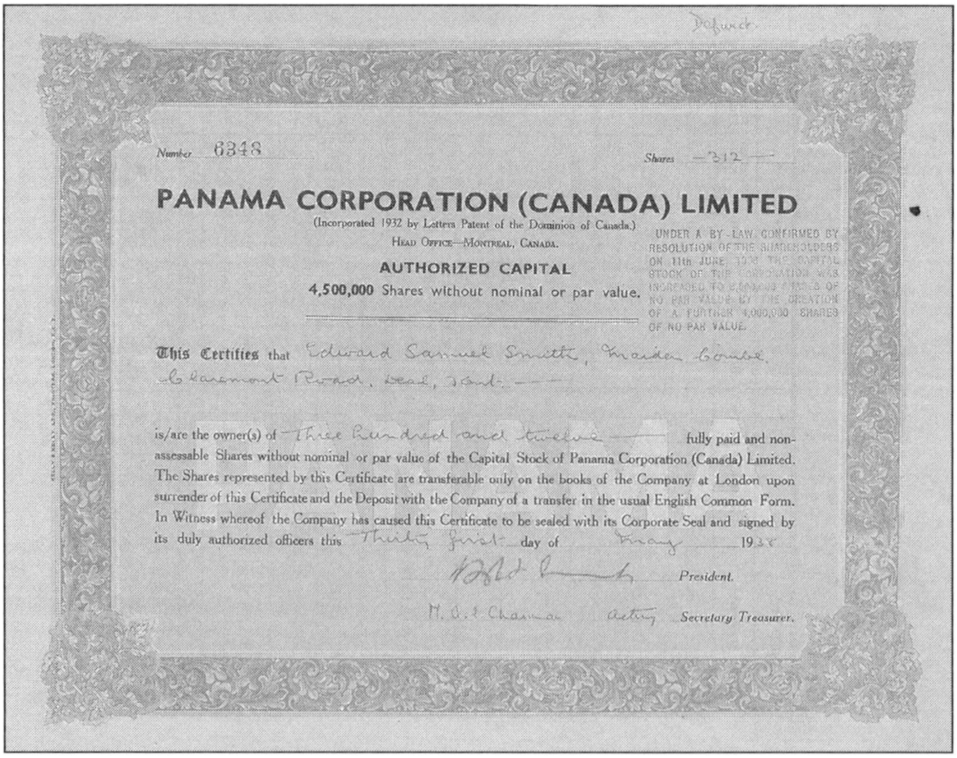

Number 6348

Shares —312—

PANAMA CORPORATION (CANADA) LIMITED

(Incorporated 1932 by Letters Patent of the Dominion of Canada.)

HEAD OFFICE—MONTREAL, CANADA.

AUTHORIZED CAPITAL

4,500,000 Shares without nominal or par value.

UNDER A BY-LAW CONFIRMED BY RESOLUTION OF THE SHAREHOLDERS ON 11th JUNE, [illegible] THE CAPITAL STOCK OF THE CORPORATION WAS INCREASED TO [illegible] SHARES OF NO PAR VALUE BY THE CREATION OF A FURTHER 4,000,000 SHARES OF NO PAR VALUE.

This Certifies that Edward Samuel Smith, Maiden Combe, Claremont Road, Deal, Kent.

is/are the owner(s) of Three Hundred and twelve fully paid and non-assessable Shares without nominal or par value of the Capital Stock of Panama Corporation (Canada) Limited. The Shares represented by this Certificate are transferable only on the books of the Company at London upon surrender of this Certificate and the Deposit with the Company of a transfer in the usual English Common Form. In Witness whereof the Company has caused this Certificate to be sealed with its Corporate Seal and signed by its duly authorized officers this Thirty first day of May 1938

[illegible] President.

M. … Chaman acting Secretary Treasurer.

British had any military and political intentions. This concern was first raised in 1926[68] by Mr. Richard Oglesby Marsh (1883-1953) who had been the Commercial Attache at the US Embassy in Panama in 1910. He returned to Panama in 1923 on behalf of Henry Ford and Harvey Firestone to explore for land suitable for rubber plantations. He became interested in the Kuna Indians, leading the ethnographic Marsh Darien Expedition of 1924-1925, and supported the Kuna or Dule (Tule) Revolution of 1925 by writing "The Declaration of Independence of the Tule People of San Blas and Darien"; for his trouble he was deported from Panama.[69] Described as a

> self-appointed guardian[s] of the Monroe Doctrine," he persuaded the Senate Foreign Relations Committee to investigate the British concessions. Marsh claimed that "the concession is a serious menace, both military and industrial, to American interests in Panama" and that the real purpose was "to block the proposed American development of rubber in Panama.

The company was criticized again in 1930 in a xenophobic article in the popular New York current affairs magazine of the day, *World's Work*, titled "Why Britishers in Panama?" by Henry Kittredge Norton (1884-1965), described as an educator, journalist, author and businessman (Norton, 1930). The article was presented to the Committee on Foreign Relations of the United States Senate at the 72nd Congress in 1932 (Norton, 1930).

Cana Exploration (1927-1932)

In August 1927 the Panama Corporation Ltd. obtained an option to purchase a large concession of over 100,000 acres (405 km^2) from "the old Darien Gold Mining Co." covering 71 mining claims for alluvial gold in the "Cana River country". The option was for three years until 1930. This was in addition to 4,600 square miles (11,914 km^2) of government concessions that it had been granted in Veraguas and Darien (Riddell, 1927). The corporation explored Cerro San José at

Cana as a bulk mineable gold target in 1927-1932 and estimated reserves of 1.8 Mt at 1 g/t Au, with potential for 11 Mt (Foster, 1939). This was the first instance of the evaluation of a bulk mineable hard-rock gold deposit in Panama, as opposed to an alluvial or gold vein-type gold deposit, a type of deposit which that would not be economic to mine for another half century until the development of heap leaching for low grade oxide gold deposits in the 1970s and the strong rise in the gold price in the 1980s.

Low (1931) described the exploration work at Cana:

> On the flat below the Holy Ghost mine and the San José hill the Anglo-French company put down several bores which showed attractive values. We have also bored this flat and part also of the San José hill. There has been a cessation of work here for the duration of the rainy season, but such tests as have been made of the auriferous alluvium tend to show that, on account of the mud which exists and the very fine and leaf-like nature of the gold, some efficient and economical system of extraction will be required if the treatment of the material is to be profitably carried out.
>
> The San José hill, which abuts on this flat, was partly worked by the Anglo-French company. This is a low-grade proposition which promises to run into large tonnages, but here again some economical method of mining and treatment will have to be adopted if success is to be attained.

Stanley Low, MIMM

The exploration manager for the Panama Corporation was V. F. Stanley Low, Member of the Institue of Mining and Metallugy, MIMM, consulting engineer. (The author is a Fellow of the same professional society today a century later.) He described the Panama exploration in two articles in The Mining Magazine (Low, 1931). The geologist was J. M. Henderson. Other named employees were Alan Gibb, cartographer and Daniel Gibson, pioneer of Bank Drilling.

Vernon Forster Stanley Low (1871-1956) was an Australian

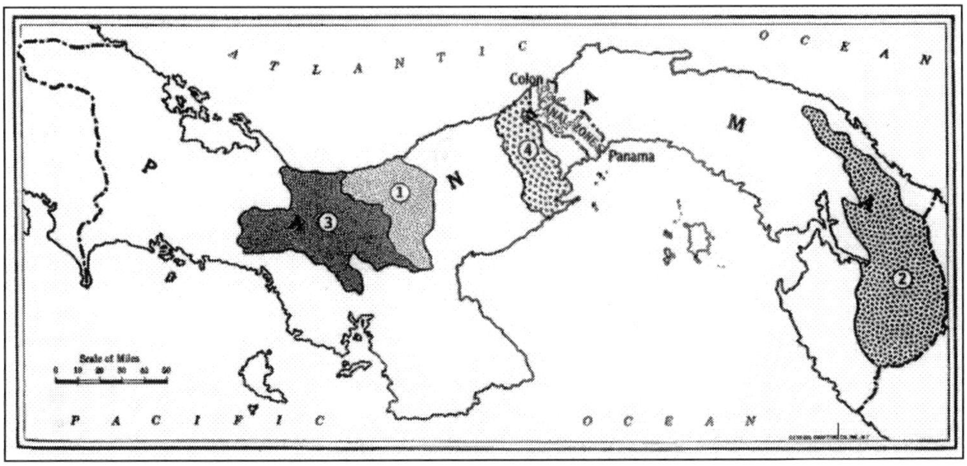

Above: The concessions of the Panama Corporation Ltd in 1930 (Norton, 1930). Key: 1) Veraguas, 1924; 2) Darien, 1925; 3) Chiriqui, 1929; 4) Rodriguez, 1930.

Below: The concessions of Panama Corporation Ltd. (Low, 1931).

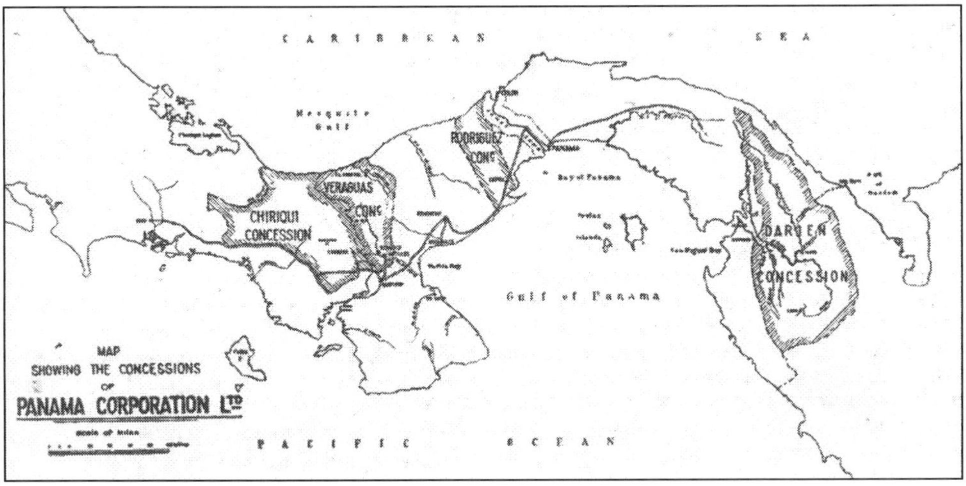

mining engineer.[70] He studied at Melbourne University and graduated BCE and MME. From 1894 to 1898 he worked for Mount Lyell Mining and Railway Co., Ltd., Tasmania, gaining practical underground experience and rising to the position of shift boss and in 1896 being made first assistant to the manager of the mining department. During 1898 Mr. Low gained further experience, as timber-man underground and later as mill-hand at the concentrator of

Stanley Low (Low, 1931).

Broken Hill Proprietary Co., Ltd., New South Wales. In the following year he was appointed general manager of Girilambone Copper Mining Co., Ltd., New South Wales.

Mr. Low became general manager of Broken Hill Proprietary Block Ten Co. in 1902, holding this position until 1911, and during this period visited the Kalgoorlie goldfields, Malaya, Indo-China, China, Japan and the Philippine Islands, and mining fields and industrial plants in Great Britain, Europe and the USA

In 1912 Mr. Low did consulting work in London and carried out investigations into Cornish tin mines, continuing these in 1913 after work connected with mines in Moscow and Siberia. He returned to Australia in 1914 for the reorganization of the Lloyd Copper Co., Ltd., Burraga, and of the Corinthian North Gold Mines, Western Australia.

In the following year he visited the Kimberley diamond mines, South Africa, and worked as mine captain of Glencairn Main Reef, Germiston. He obtained the South African Mine Manager's Certificate in 1916 and was mine captain at Crown Mines, Ltd., Johannesburg, for a short time before returning to London to enlist in H.M. Forces. He was not accepted, however, and from 1916 to 1918 worked in the Aerial Inspection Department.

After the war Mr. Low spent two years inspecting mining

and rubber properties in Kelantan and the Malay States, and in 1921 completed construction of the Lake Copper Proprietary Co., Ltd., flotation plant and reopened their mine in Sweden. He subsequently worked on reopening an oil-shale mine in Estonia, and during 1924 was joint manager of African Manganese Co., Ltd., Gold Coast. From 1925 to 1927 Mr. Low inspected gold properties in Nicaragua, manganese deposits in Greece and concessions of the Panama Corporation in Central America. He was appointed consulting engineer to the Panama Corporation in 1927 and held this position until 1932, subsequently travelling widely in connection with many new enterprises. After his retirement from professional work Mr. Low continued to travel extensively and then, in 1950, settled permanently in New South Wales. He died in Sydney in 1956 aged 85.

Duncan Elliot Alves

The Panama Corporation was managed by Duncan Elliot Alves (1870-1940), Chairman, who was born in Dunedin, New Zealand and emigrated to England in 1899 with his parents and sister.[71] They lived in Tunbridge Wells from 1905.

His father was also a mining entrepreneur. John Alves (1834-1910), who was born in Dumfries, Scotland and died at Tunbridge Wells in 1910 aged 75, although he spent most of his life in Australia and New Zealand. He founded a large engineering works at Victoria and was one of the proprietors of the Fernbull Colleries, near Dunedin, where he invented and erected an aerial railway for conveying coal from the mines to the Government railways. He designed several inventions for dredging gold and for the recovery of fine gold.[72]

In 1911 Elliot Alves married Olive Hazel Wilson (1891-1955), a Canadian who was twenty-one years his junior. They had no children. He was described as an "oil magnate", with interests in New Guinea and Venezuela. Duncan involved himself in the life of the town, especially sport. He was president of the Tunbridge Wells Football Club and of the Grove Bowling Club. At his home, a house called "The Braes" on Boyne Park, he had an extensive collection of pictures, an aviary with many rare and exotic birds and large garage which housed his collection of motor cars. Duncan was described as 'tall, debonair and nattily dressed' and as someone 'who does everything on the grand scale wherever he goes'. He earned a reputation as a generous businessman and he is remembered for the Christmas dinner he gave in 1914 in Tunbridge Wells for more than 2,500 soldiers returning from the western front.

He was the head of British Controlled Oilfields Ltd., Montreal, a company set up in 1918 that was an investment fund to acquire oilfields around the world. In 1918 it secured a huge concession of 7 million acres in Costa Rica from General Tinoco, who had recently taken power, but the deal led the company into international political controversy. The concession conflicted with existing American concessions and the two governments became involved. Britain recognized General Tinoco, while the US did not. However, another revolution brought in a pro-American government; the British concession was cancelled, and Doheny and Sinclair Oil companies were awarded the concessions instead (Norton, 1930).

Alves moved to Tidebrook Place near Mayfield, Sussex in the 1920s (address on the 1926 prospectus), and then to north Wales. He bought Bryn Bras Castle (built 1832) at Llanrug, Gwynedd in north Wales in 1920, and renovated it. He served for six years as the Mayor of Caernarvon (now Caernarfon) from 1926-1932, and as High Sheriff of the county in 1931-32. The history of the castle describes him as follows:

> Alves, who was an oil magnate, was Lord Mayor of Caernarfon for six years, and was friendly with Lloyd George and other important figures of the day. He entertained lavishly and spent a great deal of money on the house and grounds, including creating a lake ... for a private hydro-electric scheme and re-routing the northern lane where it passed closest to the castle, to enclose the small triangular paddock to the north-west.[73]

Alves died in 1940. His widow sold the castle in 1946 and she died in 1955.

Duncan Elliot Alves, Mayor of Caernarvon.[74]
Large image unknown artist © The Caernarfon Royal Town Council Collection accessed at ArtUK.org. Small image from the Kent & *Sussex Courier*, 6 th May 1910, p. 3. Both accessed at Tunbridge Tales, op. cit., with thanks to Anne Carwardine.

Hugh Marriott, MIMM[75]

Hugh Frederick Marriott (1869-1949), a Past-President of the Institution, died on 2nd March 1949, at his home in Northamptonshire at the age of 80. He was a mining geologist who was associated with the Panama Corporation, Ltd., and was president of the Panama Corporation (Canada) Ltd.

He was educated at Bishops Stortford Grammar School and from 1886 to 1891 was a student at the Royal College of Science and the Royal School of Mines, where he obtained the ARCS in Geology and the ARSM in mining.

He worked for six months on silver-lead and copper mines in the Spanish Pyrenees as assistant to Mr. P.W. Stuart-Menteath, and in May, 1892, was engaged as assistant to Mr. Hennen Jennings in the engineering department of Messrs. H. Eckstein & Co., Johannesburg, with whom he remained until his transfer in 1905 as consulting engineer to Messrs. Wernher, Beit & Co. and their successors, the Central Mining and Investment Corporation, Ltd. During this period, he was responsible for putting down the first bore-hole that proved the extension in depth of the gold-bearing reefs of the Witwatersrand.

In 1923 Mr. Marriott set up in private practice and acted as British Government representative at the Pan-Pacific Science Congress in Australia, and in the following year represented the Union of South Africa at the Empire Mining and Metallurgical Congress in London. He had held the position of Governor of the Imperial Institute. He was associated with the Panama Corporation, Ltd., and was president of its successor, Panama Corporation (Canada), Ltd. Mr. Marriott was also British Government director of Magadi Soda Co., Ltd., and Somaliland Petroleum Co., Ltd.

Walter Broadbridge, MIMM

Walter Broadbridge was a mining engineer and Director of the Panama Corporation. He died suddenly in Scotland on October 23rd, 1946, at the age of 76 as a result of a motor accident.[76]

He was a pupil and assistant under Mr. G.R. Andrews, manager of the Johannesburg Waterworks, from 1891 to 1895, and for the following eighteen months worked as assistant engineer to Mr. D.S. Johnstone, formerly manager of the George Goch Gold Mining Co., Ltd. and of Roodepoort United Main Reef Gold Mining Co., Ltd.

He then left the Transvaal for the East and spent two and a half years, from 1898 to 1901, as assistant engineer at the Redjang Lebong mines in Sumatra and the following twelve months as general manager of Kechan Gold Mining Co., Ltd., of Pahang, F.M.S. During 1902 he held the position of assistant engineer to Egyptian Soudan Exploration Co., Ltd., and from 1902 to 1904 was in the service of Messrs. Bewick, Moreing & Co. in Western Australia, first as assistant manager of the Sons of Gwalia mine, and subsequently as manager of the Vivien mine.

In 1904 he acted as consulting engineer to Messrs. L. Ehrlich & Co., for which company he visited a property in Egypt, and on his return in 1905 was appointed consulting engineer to New Districts Development Co., Johannesburg, and remained in South Africa in that capacity until April 1906.

He then became consulting engineer and technical adviser to Mr. H.M. Marshall of South African Lands and Exploration Co., and to South African Investment and Trust Co., Ltd., and reported on mines in the Witwatersrand and various parts of the Union. He and Mr. H.G. Payne went into partnership for some months during 1907 as Messrs. Broadbridge and Payne, and in June 1908. Mr. Broadbridge was appointed general manager of Prestea Block A and of Abbontiakoon Block I and other West African mines, shortly afterwards becoming also technical adviser to the group of mines at Tarkwa, under the control of Consolidated Goldfields of South Africa. In 1911 he transferred to the employment of Minerals Separation, Ltd.

During the 1914-1918 war he served as captain in the Royal Field Artillery, and on being invalided out of the Army worked at the Ministry of Munitions as Director of Aluminium Supplies, Graphite and

Carbide. He later rejoined Minerals Separation, Ltd., of which he became a Director, and was associated with Rhodesian Congo Border Concession, Ltd. He and Mr. Preston K. Horner joined in practice in 1930 and were connected with the operations of British American Mines, Ltd., in Western Australia and Siam. He became a Director of Boulder Perseverance, Ltd., in 1931.

12. *Exploration (1958 - 1996)*

There appears to have been no exploration or mining activity at Cana for three decades from 1932 until 1958. Why the hiatus and why this date? The start of exploration in 1958 coincides with the period of modern minerals exploration in Panama which took place between 1956 and 1982 (Redwood, 2021b). The Cana deposit was explored for placer gold, epithermal gold and porphyry copper deposits in the period between 1958-1996 by several companies including the South American Gold & Platinum Company, Hochschild, TexasGulf, Freeport Exploration Company, St. Joe Minerals, Gold Fields Ltd of South Africa, and Link Mineral Ventures.

Louis Martinz (1950s)

Louis Martinz (1940-1994) acquired the Cana concessions in the 1950s. He carried out no work on the mine. His assistant was Fernando Manfredo, former Administrator of the Panama Canal. Martinz was Austrian who became a Panamanian citizen. Tower described him as having "very substantial business interests in Panama, including a contracting company and has just opened a bank: he is reputed to be the wealthiest man in the Republic."[77] He was the Austrian consul, the Chief of Protocol for President Arnulfo Arias until 1968, was exiled by the military government to New York for ten years until 1978, then on his return to Panama he published a newspaper opposed to the military government, and finally was spokesman for President Guillermo Endara in 1989.[78] Mendez (2004, p. 414) reported that the mines then belonged to Martinz' heirs and were being worked by a company called Floralina.

South American Gold & Platinum Company (1958)

The South American Gold & Platinum Company (SAG&PC) examined the placer deposits of the Cana Plateau in 1958 and tried to make an agreement to carry out drilling but an American promotional group was involved and negotiated an unsatisfactory contract with the government.[79] The work was managed by William H. Breeding and a site visit was made by Arturo Betancourt (referred to by Tower as Mr. Bukofzer) and Joe de Deaux [de la Guardia named also]. They described the buildings as "being largely covered by jungle growth, the old railway to Chupe dismantled and the trail partly blocked, requiring a two day trip to cover 12 miles."[80] The company thought that some of the smaller Chocó or Pato dredges might be used there and could probably be floated there. Breeding considered "The Darien Area is one of the better potential gold placer, and possibly lode, areas in Latin America."

The South American Gold and Platinum Company (SAG&PC) of New York, founded in 1915, operated gold and platinum dredges in the Rivers Chocó, San Juan, Pato and Nechi in Colombia. It merged with International Mining Corporation (IMC) of New York in 1963.

Mauricio Hochschild Mining Group (1963-1966)

The Mintrade group owned by the Mauricio Hochschild mining group carried out exploration and relocated the old mine workings in 1963-1966. A report by George W. Tower to Mauricio Hochschild in January 1964 recommended forming a syndicate between Hochschild, SAG&PC and Thayer Lindsley.[81]

George Tower wrote in 1964[82] that the reports he consulted regarding the mine were: 1) San José Hill by Donald Foster, March 1929, 2) the Kimball report for T. Arias (no date), 3) notes of interviews with former miner T. Maslia in 1927, 4) a report by J. McClelland Henderson of Henderson, Weakes and Gibb in 1937, described as "the best report", 5) a synopsis by C. P. C. Beresford of April 3 1937, mostly quoting Henderson. Beresford concluded that:

1. There was "undoubted proofs" of the production of large quantities of gold.
2. There is a "very reasonable expectation" of finding additional high grade pipes like the South Mine, as these pipes occur in clusters according to Mr. Henderson. He also considered that insufficient work had been carried out the North Mine to determine whether there is a similar high grade zone at greater depth.
3. San José Hill is a target with a large tonnage and low grade "worthy of intensive study and prospecting".
4. The Cana Plateau has potential for alluvial gold deposits.

Texas Gulf Inc. (1970s)

In the 1970s exploration was carried out at Cana by TexasGulf Inc., which was exploring Cerro Colorado at the same time in 1975-1980, and St. Joe Minerals. This included 8,400 m of diamond drilling and 115 m of churn drilling in the area of the old mines (Ruiz-Williams, 1990). According to Nelson (1995), TexasGulf Sulfur drilled 16 holes on Cerro San José.

IXTAPA (1979-1993)

The Sociedad de Inversiones de IXTAPA S.A., a company incorporated in Panama on 4 July 1979, obtained a concession of 15,732 hectares at Cana in 1980 under the Torrijos government and in 1985 it obtained the exploitation rights valid for 60 years (Ruiz-Williams, 1990; Duran, 1998). The company was owned by Gabriel Lewis Galindo (1929-1996),[83] a Panamanian businessman and diplomat, who was the Ambassador to the US in 1977-1978 during the negotiation of the canal treaties. It was inherited in 1996 by his sons Eduardo and Fernando Lewis Galindo. IXTAPA had a joint venture with Freeport in 1985-1988 (Ruiz-Williams, 1990). Gold Fields Limited of South Africa evaluated the project in 1992.

Freeport Exploration Company (1985-1988)

Freeport Exploration Company, a subsidiary of Freeport-McMoRan Copper & Gold Inc, New Orleans (now Freeport-McMoRan Inc.), explored for copper and gold in Panama from 1985 to 1990 and left after the US invasion. Freeport, a Gulf sulfur company, discovered the Ertzberg copper-gold skarn deposit in similar but more extreme climate and mountainous, jungle terrain in Irian Jaya, Indonesia in 1967 and began mine production in 1972 (Wilson, 1981; Mealey, 1996). Exploration near the mine led to the discovery of the giant Grasberg copper-gold porphyry only 2 km away in 1988. As a result, Freeport is now the biggest copper mining company in the world.

At the same time as Freeport was exploring in Indonesia, Freeport de Panama explored Cana in a joint venture agreement with IXTAPA in 1985-1988 (Tippett & Trever, 1989). It drilled 4,600 m in 40 holes at the North Mine and Cerro San José and defined proven and probable reserves of 268,000 tons grading 0.4 oz/t (243,125 metric tons at 13.7 g/t Au) containing 108,000 ounces (Ruiz-Williams, 1990; Nelson, 1995). It was hoping to discover a bigger deposit and did not carry out any more work at Cana.

Gold Fields Limited of South Africa Ltd (1992)

Gold Fields Limited of South Africa evaluated the project in the early 1990s but did not make an agreement with IXTAPA and did not carry out any new exploration.[84] The company carried out a 1,200 hole soil auguring program in 1992 and defined several soil anomalies >1 ppm Au (Nelson, 1995).

ANCON (1993-1996)

In 1993 the Panamanian environmental non-governmental organization (NGO) ANCON (Asociación Nacional para la Conservación de la Naturaleza) bought 93% of IXTAPA for $125,000, the Lewis Navarro brothers retaining 3%, in a bizarre story told by Pastor E. Durán Espina titled *¿Es Verde el Color del Oro?* (*Is Green the Colour of Gold?*) (Durán, 1988) The NGO's plan was to cancel the mining concession and thereby eliminate mining, but they were informed by the mining authority DGRM (General Department of Mineral Resources) that there was a queue of mining companies waiting for them to do just that in order to stake a new mining concession. ANCON discovered that to prevent mining it had to comply with a series of requisites stipulated by the Mining Law in order to maintain the mining concession valid, and it had to carry out an exploration program designed to lead to the exploitation of the mine, contrary to its ideals and its legal status as a not-for-profit organization.

Link Mineral Ventures Limited (1995-1997)

Accordingly, IXTAPA formed a joint venture with the junior mining company Link Mineral Ventures Ltd of Toronto in 1995-1996. The President of Link Minerals was Dr. John Hansuld (1931-2019), a highly respected Canadian geochemist, mine-finder, entrepreneur, company-builder and financial-leader, former Chief Geochemist and Vice President of Amax Inc. (1961-1983), President and CEO of Canamax Resources (1983-1989), a founder of the Association of Applied Geochemists (1970), President of the Prospectors and Developers Association of Canada (1993-1996), and was inducted into the Canadian Mining Hall of Fame in 2012.[85] Another director was Peter Bojtos, an old Panama hand who had been Chairman and CEO of Greenstone Resources Ltd (1993-1995). Link Minerals stated in its 1996 Annual Report that it had fulfilled the conditions to earn 80% of the Cana project but that there was a dispute with IXTAPA. Their deferred exploration expenditure at 31 December 1996 were US$2.270 million. Link continued the legal dispute in 1997 but then wrote off its loss and closed its office in Panama City. Meanwhile, in 1996 during the dispute, the Panamanian government banned all mining activity within the Darien National Park.

Darien National Park

The Alto Darién Protection Forest was created in 1972 and became the Darién National Park in 1980, with an area of 579,000 hectares (5,790 km^2). It was designated as a UNESCO biosphere reserve in 1983. Mining was banned in the Darien National Park in 1996 which ANCON took credit for. The Cana exploration camp and grass airstrip were subsequently used as a jungle lodge for ecotourism and ornithologists.

José Oller Collection

The Espiritu Santo de Cana mine camp.

Above: The village of Boca de Cupé, District of Pinogana on the Tuira River was the head of river navigation and the terminus of the trail to the Cana gold mine.

Below: The entrance to the Cana mine camp was on the Cana River.

Above: Puente del Pueblo over the Cana River
and the Campamento del Pueblo for workers.

Below: The general manager's house
and the director's guest house at the Cana mine.

Above: Houses at the Cana mine camp.

Above: The Cana mine camp.

Below: The railway station in Cana after 1912.
The narrow gauge railway was powered by *tracción de sangre* (mules).

Above: Visit to the Cana mine by the Bishop of Panama.
Seen with mule-drawn carriage on the narrow gauge railway.

Below: The commissary building with West Indian men and children,
but no women.

Above: The medical clinic and dispensary (*botica*)
and Dr. S. Gibbs with two children.

Below: A herd of dairy and beef cattle supplied the Cana mine camp.

Above: Mules used for transport between the Cana mine and Boca de Cupé on the Tuira River.

Below: Troughs (*pesebreras*) and stream to feed and water the mules.

Above: Don Jil Francisco Sanchez, a mine supervisor of the Darien Gold Mining Company, ready to ride to the port of El Real.

Below: A view of the Cana mine camp.

Above: Mr. Horace O. Blake,
chief accountant of the Administrative Office at the Cana mine.

Below: An armed detachment of force of public order at Cana provided by the Panamanian government under the command of the *Corregidor* (Mayor).

Above: Four men bathing at a bathing pool below a waterfall in the headwaters of the Seteganti River, a tributary of the Tuira River.

Below: Sres. Miguel Abrahams and Atilio Repetto, a sub-contractor of the company, riding to the Escucha Ruido station.

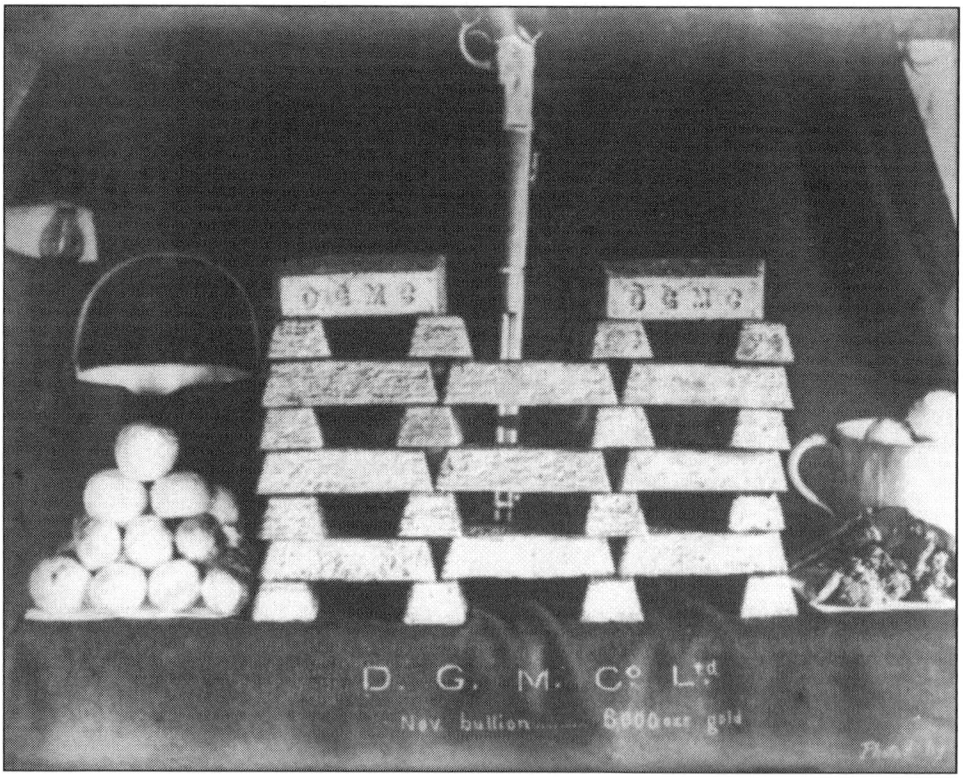
D. G. M. Co Ltd
Nov. bullion ...... 6000 ozs gold

Above: Spanish 17th century wooden tread-wheel used to remove water from the mine. Five were found preserved under water in the Espiritu Santo de Cana mine.

Left above: Expat employees posing on an improvised rope bridge across the Escucha Ruido River on the Cana Plateau.

Left below: The production of the Espiritu Santo de Cana mine in November 1904: 27 gold ingots weighing 6,000 ounces. At the time, it was worth about USD 0.75 million, today about USD 27 million.

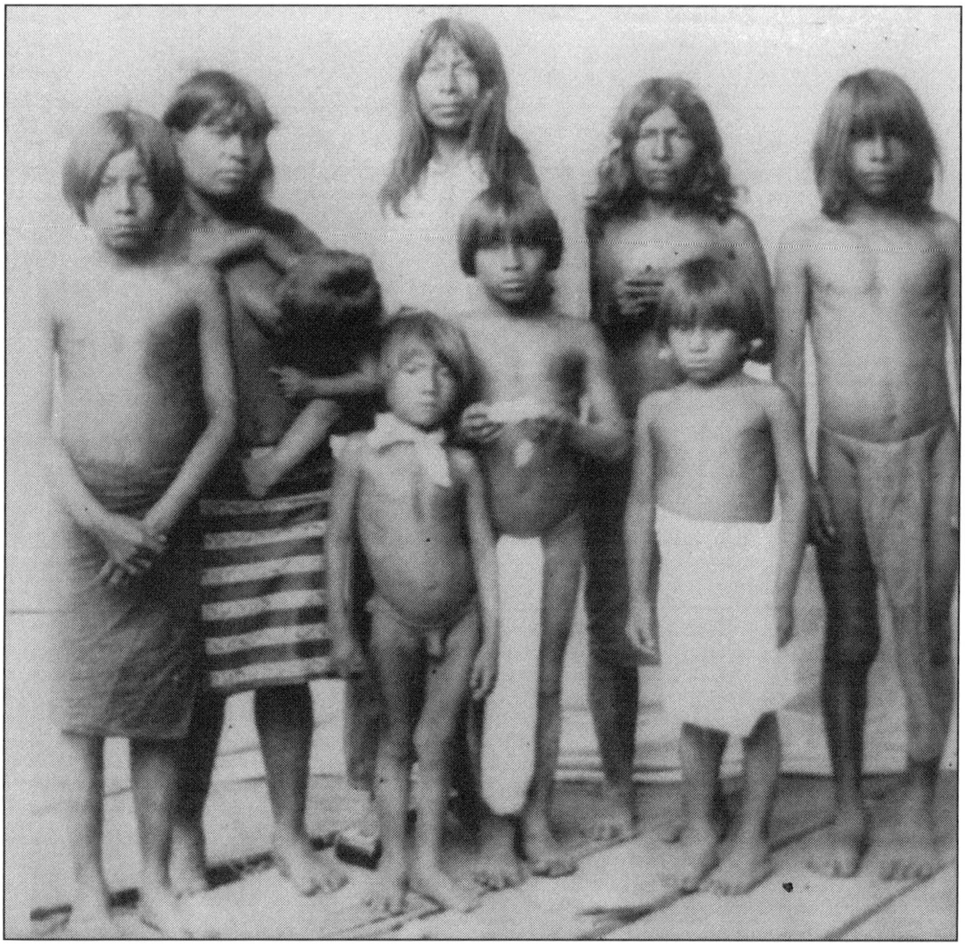

Above: Chocó (Embera) family.

Right: Chocó (Embera) Indian.

Above: The Forty-Stamp mill at the Espiritu Santo de Cana gold mine.

Below: The mill and processing plant at the Espiritu Santo de Cana gold mine.

Rear view of the Forty-Stamp mill.

Old compressor plant building.

Machine shop with belt-driven lathe.

Above: Mine cars at the Cana mine.

Below: A bank of six Wilfey tables (invented 1895) used
to concentrate gold by gravimetry using running water and vibration.

Above: A belt-driven dynamo and electric pump,
probably used for placer mining at the Cana mine.

Below: Water pipe being installed from the pump house, San José de Cana.

Above: Building a channel or aqueduct for placer mining at San José de Cana.

Below: Placer gold mining was carried after 1908 by the French when they managed the Darien Gold Mining Company at San José where the Cana River flows into the Cana Plateau, about 1 km east of the mine shafts that were in the mountains.

Above: Placer gold mining using monitors (high pressure water canon) at San José de Cana.

Below: Aqueduct to transport placer material to the washing plant, San José de Cana.

Above: Water pipes and raised aqueduct at the San José de Cana placer mine.

Above: Washing plant for placer gold at San José in the Cana Plateau. Material is transported to it in a raised channel. Washed gravel exits over the slide to the pile in the foreground. It is described in Spanish as *un terminal de acequia*.

Below: Crushing plant with feed on left by mine cars on elevated railway.

Above: Ore transport by electric locomotive and mine cars.

Right above: Loaded mine cars being pushed manually at the San José de Cana placer mine.

Right below: Placer gravel being mined and transported by mine cars at the San José de Cana mine.

Above: Group of mine managers at Cana.

Above: Two mine managers at an aqueduct.

Below: Group of employees with a child at Cana.

Above: Group photo of Cana employees celebrating independence day from Colombia on 3 November, with Panamanian and US flags.

Below: Group photo with dogs at the Cana mine.

Above: Group photo with at least five dogs and a pet spider monkey on a holiday, probably celebrating independence day from Colombia on 3 November with the flags of Panama, the United Kingdom and the USA.

Below: Group of expat mine managers at Cana with the chief cook "Ceff."

Vicente Pascual Collection

Photo postcard of Cana mine camp, 1906.

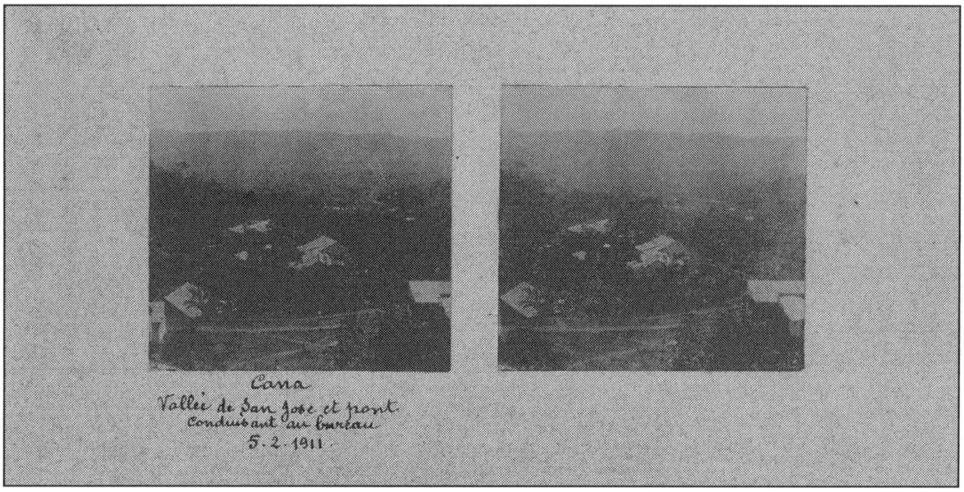

A stereo photo pair of a French director's visit.

Cana. San José Valley and bridge leading to the office,
5 February 1911

Cana.
San José Valley and bridge
leading to the office,
5 February 1911
(From a stereo photo pair of a French director's visit).

Cana.
On the railway,
Mt. Kitchner.
M. Brerault, M. Lue, M. Audemar,
1 February 1911
(From a stereo photo pair of a French director's visit).

Cana.
Team of oxen, Escucha Ruido,
25 January 1911
(From a stereo photo pair of a French director's visit).

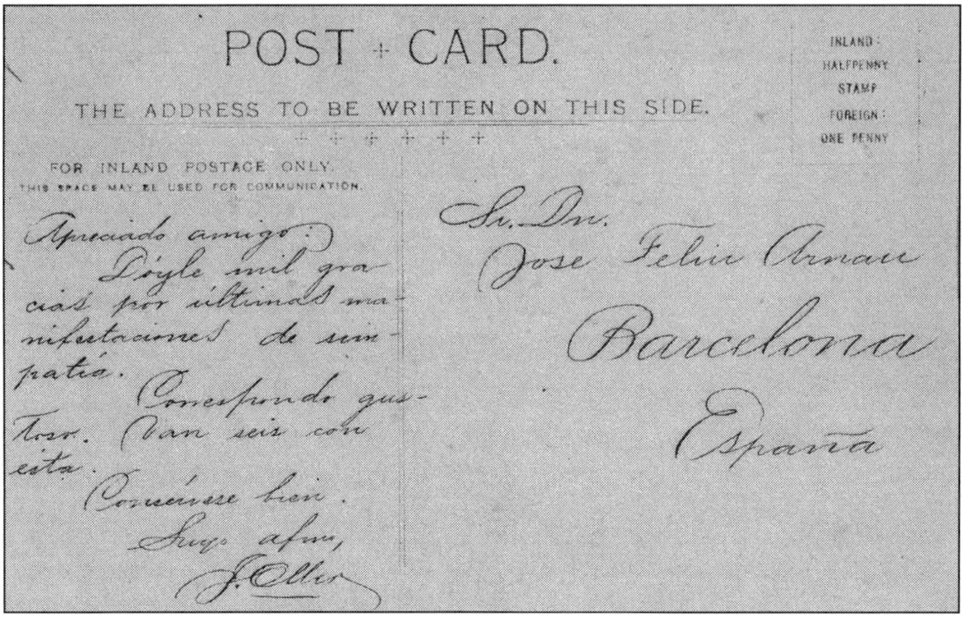

POST + CARD.

THE ADDRESS TO BE WRITTEN ON THIS SIDE.

INLAND: HALFPENNY STAMP FOREIGN: ONE PENNY

FOR INLAND POSTAGE ONLY.
THIS SPACE MAY BE USED FOR COMMUNICATION.

Apreciado amigo:
Dóyle mil gra-
cias por ultimas ma-
nifestaciones de sim-
patía.
Correspondo gus-
toso. Van seis con
esta.
Consérvese bien.
Suyo afmo,
J. Oller

Sr. Dn.
Jose Felix Arnau
Barcelona
España

Reverse side of photo postcard of Cana mine camp.
Message written by José Oller
to Don José Felix Arnau, Barcelona, Spain,
24 September 1906.

Dear friend,

Thank you very much for your recent expressions of friendship.

I am happy to reciprocate. That makes six with this one.

Take care of yourself.

Yours affectionately,

J. Oller

Kevin Heather Collection: April 1987

Cana airstrip.
View from the camp as a Britten-Norman HP2 Islander aircraft HP-1016 lands.

Above: Cana exploration camp from the airstrip.

Below: Cana airstrip.

Above: Cana exploration camp.

Below: Cana airport terminal.

Old mining machinery at Cana.

Triple Chilean or arrastra mill at Cana.
Kevin Heather on old mining machinery.

Above: Old mining machinery at Cana.

Below: A steam-powered pumping engine
manufactured at Fraser & Chalmers, Limited, at Erith, Kent, England.

Above: A 20 hp steam locomotive manufactured by Orenstein & Koppel of Drewitz, Potsdam, Germany. It was used on a 47.5 km long, 0.60 m narrow gauge railway, from Boca de Cupé on the Tuira River to the mine built in 1907.

Below: Kevin Heather at Cana mine entrance with date inscribed: 1905.

Above: Old mining machinery at Cana.

Below: Greg Maynard, Freeport geologist at Cana.

Bert Harrison Collection

Orchestra formed by expat mine employees, circa 1900.

Above: Escort ready to ride from Cana to Boca de Cupé with the mail and the bullion Woakes, (1923) captioned the photo as American employees leaving camp when the US annexed the Canal Zone. Actually, employees did leave in 1903 on rumors that the Colombian army would invade the new republic at the nearby frontier.

Below: Man with butterfly net at a waterfall in the headwaters of the Seteganti River, a tributary of the Tuira River, circa 1900.

Above: A group of Embera women and children with three expat mine employees, probably medical staff, at Cana, circa 1900.

Below: Bert Harrison on horseback with an Embera man and a blue and gold macaw, circa 1900.

Above: Bert Harrison.

Above: A pet sloth with a string around its waist to stop it from running off!
Circa 1900.

Above: Padre with kitten, circa 1900.

Left above: Heenan Engine Shaft (later Maisonabe) and Forty Stamp mill, Espiritu Santo Cana gold mine, circa 1900.

Left below: Pack mules at Cana, circa 1900.

Above: Wilfey tables for the concentration of gold
in the stamp mill at Cana, circa 1900.

Above: The railway station in Cana. Note two tracks, circa 1900.

Below: Horses and wagons at Cana, circa 1900.

Above: The Heenan Engine Shaft, Espiritu Santo de Cana mine, circa 1900.

Historical Collection

Ore specimens, Espiritu Santo de Cana (Woakes, 1899).

Heenan Engine Shaft, Espiritu Santo mine, Cana (Woakes, 1899).

Below: Stamp Mill, Espiritu Santo mine, Cana (Woakes, 1899).

Above: Cana, from the high level mill ditch (Woakes, 1899).

Above: View of the Cana camp in 1912, taken by zoologist Edward Goldman.

Above: Native houses at Boca de Cupé, on the Tuyra River (Woakes, 1923).

Below: A funeral of a West Indian native passing the Company's hospital (Woakes, 1923).

Above: El Palo camp, Darien Concession,
the limit of river transport
(Letcher, 1930).

Below: Drilling to evaluate alluvial gold deposits
on the Balsa River, Darien Concession
(Letcher, 1930).

Above: Light steam locomotive from the narrow gauge railway from Boca de Cupé to the mine built in 1907. It was manufactured by Orenstein & Koppel of Germany at Drewitz in Potsdam, near Berlin, a specialist in light railway locomotives. This engine is 30 - 40 hp. Below: This is a smaller engine of 20 hp (G. Dawson, 2009).

Above: A steam-powered pumping engine manufactured by Fraser & Chalmers, Limited, at Erith, Kent, England (G. Dawson, 2009).

Below: Advertisement by Fraser & Chalmers for a similar steam pumping engine.

Afterword

The Cana area has been de-populated and left to the illegal activities of Colombian drug traffickers, human smugglers, terrorists, and artisanal gold miners. The former Cana gold mine, once the largest in Panama, ought to be a site of historical importance (patrimonio cultural) given its history and the mining remains of major industrial archaeological interest of late 19th to early 20th century British and European mining equipment including two narrow gauge steam locomotives, rail coaches and mine cars, crushers, pumps and other artefacts, now rusting in the rain forest. The mines themselves now lie overgrown and hidden in the dense second growth rain forest. The ruins of the once important town of Santa Cruz de Cana are hidden in the jungle – the traces of the town would show up well and could be mapped out if any archaeologist cared to fly a Lidar survey of the type used to discover hidden Mayan cities in the Yucatan with spectacular effect. The town of Cana had about 900 houses and a church in the 17th century, and in the early 20th century it had a population of about 800 people and 180 houses. The community of Panamanians and West Indians with British, French and American engineers and administrators is long since gone, and this important site of Panama's historical heritage is forgotten.

Acknowledgments

The author would like to thank Germinal Sarasqueta Oller, Panama for giving me a copy of his grandfather's book (Oller, 1975), and for lending me the negatives of his grandfather Jose Oller's collection of photos, and Peter Tristram, Australia, for sending me copies of the photos of the mine that belonged to his great uncle, Bert Harrison, and to Carla Black, Volcán, and Charlotte Elton, Panama, for the introduction. The photos were actually taken by Leo George Inglis, Cashier at the mine, who was a "photographer in his spare time" (Oller, 1975, p. 14) and gave or sold copies to various people. His photos have appeared in publications and other collections. For the Oller Collection, I scanned the 120 sized negatives, some of which have heat damage, which are copies made of prints rather than original negatives. Vicente Pascual, David S. Plowman and Nicolás Liakópulos Falcón are also thanked for sharing their knowledge and permission to use images of photos, tokens and stock certificates from their collections.

Thanks are due also to Dr. Glyn Dawson, Chicago, for permission to publish his photos of the steam locomotives, and to Jeremy Martin, London for identifying the locomotives. Several geologists who visited the project in the 1960s to 1980s are thanked for sharing their recollections, including Gerard Ellis, former Director of Mines, Dominican Republic, who worked for Hochschild in the 1960s and visited Freeport in 1985; Raymond Jannas, who worked for Gold Fields Ltd; Kevin Heather who visited the Freeport exploration in 1987; and Carl Nelson who kindly shared photos also. Frances Perry and Linda Dawes of the IOM3 Library in London and later Grantham found many old references to Darien in the mining press. In Panama, Ayleen Cedeño is thanked for searching the National Archive and Tómas Mendizabal for sharing historical maps.

Sources

Archives

IOM3: The Institute of Materials, Minerals & Mining Library, London and Grantham, England.

Northern Mine Research Society, England.

- Obituaries of Members of the Institute of Mining and Metallurgy (IMM, now IOM3).

ANP: The National Archives, Panama

AGI: The General Archive of the Indies, Seville

- The Darien Gold Mining Co. Ltd.

TNA: The National Archives, Kew, London, United Kingdom.

- BT: The Board of Trade company registry.
- The Darien Gold Mining Co. Ltd., 1887.
- The Darien Gold Mining Co. Ltd., 1891.
- The Darien Gold Mining Co. Ltd., 1907.
- Panama Corporation, Limited, 1926. BT 31/32749/213042
- FO: Foreign Office.

UOW: University of Wyoming, American Heritage Center

- Thayer Lindsley Papers, 1887-1977:
  Central America – Panama Gold – Cana Mines, 1964
  (George W. Tower, Mauricio Hochschild).
  Collection 6034, Box 110, Folder 1.
- Panama, Panama Corp., 1939-1944.
  Collection 6034, Box 204, Folder 1.

US Navy History and Heritage Command.

NARA: US National Archives and Records Administration.

References

- Anderson, C. L. G., 1911. *Old Panama and Castilla del Oro*. Press of the Subwarth Company, Washington, D.C, 557 p.
- Araúz, Amado, 1962. "El pasado esplendoroso de Cana" (Entrevista a Medardo Murillo). Panamá, *El Panamá America*, junio 11 de 1962. (cited by Araúz, 2005b).
- Araúz, Amado, 2005b. "Imágenes de Santa Cruz de Cana." *Epocas*, Segunda Era, *La Prensa*, Panama, Año 20, No. 11, Noviembre 2005, p. 2-3. (Extracts from book of visit by the French shareholders' lawyer Albert Fraissignes in 1901, and interview by Arauz of an old miner, Medardo Murillo, in 1962)
- Ariza, Andrés, 1774. "Comentarios de la rica y fertilísima provincia del Darién, 1774." In Baquero, Á. & Vidal, A. (eds). *La gobernación del Darién a finales del siglo XVIII. El informe de un funcionario ilustrado*. Uninorte, Barranquilla, 2004.
- Barnhart, J. H., 1945. Robert S. Williams. *Journal of the New York Botanical Garden*. Vol. 46, No. 546, p. 146-147.
- Blas Aritio, L., 2014. *Vasco Núñez de Balboa, La Crónica de los Cronistas de Indias*. Ediciones Balboa, Panamá, 423 p.
- Bray, W., Cooke, R. H. & Redwood, S. D., 2021. "Early Metalwork in Caribbean Colombia and Southern Central America." In: McEwan, C. & Hoopes, J. W., eds. *Pre-Columbian Art from Central American and Colombia at Dumbarton Oaks*. Dumbarton Oaks Research Library and Collection, Trustees for Harvard University, Washington, D. C., p. 541-567 and Bibliography p. 713-745.
- Bush, M.B. & Colinvaux, P.A., 1994. "Tropical Forest Disturbance: Pa-

leoecological Records from Darien, Panama." *Ecology*, Vol. 75, No. 6, p. 1761-1768.

• Castillero Calvo, A., 2019a. Conflictos sociales, guerra, piratas y Pax Hispana. In, Castillero Calvo, A. (ed.), *Nueva historia general de Panamá*. Comisión 500 Años de Fundación de la Ciudad de Panamá, Panama and Phoenix Design Aid A/S, Denmark., 6 volumes. Vol. I, Part 2, p. 785-832.

• Castillero Calvo, A., 2019b. Las fuerzas regulares y las milicias. In, Castillero Calvo, A. (ed.), *Nueva historia general de Panamá*. Comisión 500 Años de Fundación de la Ciudad de Panamá, Panama and Phoenix Design Aid A/S, Denmark. 6 volumes.Vol. I, Part 2, p. 833-876.

• Castillero Calvo, A., 2019c. Cronología Contextualizada para la Nueva historia general de Panamá. In, Castillero Calvo, A. (ed.), *Nueva historia general de Panamá*, Comisión 500 Años de Fundación de la Ciudad de Panamá, Panama and Phoenix Design Aid A/S, Denmark. 6 volumes. Vol. 3, Part 2, p. 1091-1154.

• Catat, L., 1889. "Les habitants du Darien meridional." *Revue d'ethnographie*, Vol. 7, p. 397-421.

• Chase A.B., 1907. "Caving system at the Darien Mine, near Cana, Panama." *Mining and Scientific Press*, San Francisco, August 24, 1907, p. 238.

• Chibas, E. J., 1898. "The Darien Gold-Mining Region in Colombia." *Engineering Magazine*, October 1898, p. 49-63.

• Craig, J. R. & Rimstidt, J. D., 1998. "Gold production history of the United States." *Ore Geology Reviews*, Vol. 13, p. 407-464.

• Cuestas Gómez, Carlos Humberto, 1999. *Panamá y Costa Rica, entre la diplomacia y la guerra*. Litho Editorial Chen, S.A., Panama, 420 p.

• Cullen, E., 1852. *Isthmus of Darien Ship Canal*. Effingham Wilson, London, 1st edition, 1852, 64 p.

• Cullen, E., 1853. *Isthmus of Darien Ship Canal; With a Full History of the Scotch Colony of Darien, Several Maps, Views of the Country, and Original Documents*. Effingham Wilson, London, 2nd edition, 1853, xii + 204 p.

• Dalrymple, Sir J., 1778. *Memoirs of Great Britain and Ireland*. Printed for A. Strahan & T. Cadell, The Strand, [London] and J. Bell, W. Creech & E. Balfour, Edinburgh, 3 volumes, 1771, 1773, 1778. New edition 1790. Vol III, Part III, Book VI, p 121-171, Adventures of Paterson, and Description of Darien.

• Dampier, W., 1697. *A New Voyage Round the World.* James Knapton, London, 1697, [15] + 550 p. Hummingbird Press, London, 1998, x + 294 p.

• De Banville, M., 2005. *Canal Francés: La aventura de los franceses en Panamá.* Ediciones Canal Valley S.A., Panama, 171 p.

• Degoutin, M., 1913. *Auriferous Andesite Mines.* The Darien Gold Mining Co. Ltd., Manchester. Printed by Taylor, Gannett, Evans & Co. Ltd., Manchester and Reddish, February 1913, 26 p. (cited by Oller, 1933).

• De La Guardia de Corró, M. M., 2009. *Gabriel Lewis Galindo: Hasta la última gota.* Cargraphics S.A., Cali, Colombia, xiii + 357 p.

• Durán, E., 1998. "Polémica Sobre las Minas de Cana: Es Verde el Color del Oro?" *La Prensa*, 30 July 1998. http://www.ambiente-ecologico.com/revist50/duran50.htm.

• Esquemeling, J., 1685/1911. *The Buccaneers of America.* William Crooke, London, 1685. George Allen & Company, Ltd., London, 1911, 502 p. The Second Volume: Part IV. *Containing the Dangerous Voyage and Bold Attempts of Captain Bartholomew Sharp, and others; performed upon the Coasts of the South Sea, for the space of two years, &c. From the Original Journal of the said Voyage written by Mr. Basil Ringrose.* 1911, p. 272-502.

• Forrester, A., 2004. *The Man Who Saw the Future: William Paterson's Vision of Free Trade.* Thomson Texere, New York, xviii + 372 p.

• Foster, D. F., 1939. *Report on San José Hill, Darien, R. de Panama.* Report for Panama Corporation (Canada) Ltd., University of Wyoming, American Heritage Center, Thayer Lindslay Papers, Collection 6034, Box 204, Folder 1, 7 p.

• Fraissignes, A., 1902. *Excursion d'un actionnaire aux Mines de Cana.* Paris, janvier 1902. (cited by Aráuz, 2005a, b).

• Gisborne, L., 1853. *The Isthmus of Darien in 1852. Journal of the Expedition of Inquiry for the Junction of the Atlantic and Pacific Oceans.* Saunders and Stanford, London, v + 238 p.

• Goldman, E.A., 1912. "Completion of the Biological Survey of the Panama Canal Zone. Explorations and Field-Work of the Smithsonian Institution in 1912." *Smithsonian Miscellaneous Collections*, Vol. 60, No. 30, p. xxx.

- Goldman, E.A., 1920. "Mammals of Panama." *Smithsonian Miscellaneous Collections*, Vol. 69, No. 5, 309 p., 39 plates.
- Heckadon-Moreno, S., 2006. "Exploracíones botánicas de R. S. Williams en Coclé y Darién (1908)," p. 145-159 in *Selvas entre dos mares: Expediciones cientificas al Istmo de Panama, siglos XVIII-XX*. Smithsonian Tropical Research Institute, Panama, 312 p.
- Heckadon-Moreno, S., 2004c. "Seth E. Meek and Samuel F. Hildebrand: The fishes of Panama." In Heckadon-Moreno, S., 2004a, Chapter 14, p. 156-169.
- Heckadon-Moreno, S., 2006. *Selvas entre dos mares: Expediciones cientificas al Istmo de Panama, siglos XVIII-XX*. Smithsonian Tropical Research Institute, Panama, 312 p.
- Heckadon-Moreno, S., 2006. "Exploracíones botánicas de R. S. Williams en Coclé y Darién (1908)," in Heckadon-Moreno, 2006, p. 145-159.
- Hull, C.F.H., 1940. *Panama Corporation (Canada) Limited*. Report for Panama Corporation (Canada) Ltd., University of Wyoming, American Heritage Center, Thayer Lindslay Papers, Box 204, Folder 1, 8 p.
- Joyce, L.E.E. (Ed.), 1934. *A New Voyage and Description of the Isthmus of America by Lionel Wafer, Surgeon on Buccaneering Expeditions in Darien, the West Indies, and the Pacific from 1680 to 1688, With Wafer's Secret Report (1698), and Davis's Expedition to the Gold Mines (1704)*. Hakluyt Society, Oxford, 1934. Lxxi + 221 p.
- Letcher, O., 1930. "Gold in Panama. Engineering and Mining World," Vol. 1, No. 3, March 1930, p. 153-154.
- Low, V. F. S., 1931. "Panama." *The Mining Magazine*, Vol. 44, April 1931, p. 201-209; May 1931, p. 273-279.
- Masefield, J., 1906. *On the Spanish Main: Or, some English forays on the Isthmus of Darien, with a description of the buccaneers and a short account of old-time ships and sailors*. Methuen & Co., London, [18], 344 p.
- McCullogh, D., 1977. *The Path Between the Seas: The creation of the Panama Canal, 1870-1914*. Simon & Schuster, New York, 698 p.
- Mealey, G.A., 1996. *Grasberg: Mining the Richest and Most Remote Deposit of Copper and Gold in the World, in the Mountains of Irian Jaya, Indonesia*. Freeport-McMoran Copper & Gold, New Orleans, 384 p.
- Meek, S. E. & Hildebrand, S. F., 1916. "The Fishes of the Fresh Waters

of Panama." *Field Museum of Natural History, Publications, Zoological Series*, December 28, 1916, Vol. X, No. 15, p. 217-374.

- Mena García, C., 2011. *El oro del Darién: Entradas y cabalgadas en la conquista de Terra Firme (1509-1525).* Fundación Pública Andaluza, Centro de Estudios Andaluces, Seville and Consejo Superior de Investigaciones Científicas, Madrid, 640 p.
- Méndez, T. E., 2004. *El Darién: Imagen y Proyecciones.* Instituto Nacional de Cultura (INAC), Panamá, Panamá. 1st edition 1979, 2nd edition, 2004, 560 p.
- Nelson, C. E., 1995. "Porphyry copper deposits of southern Central America." *Arizona Geological Society Digest*, vol. 20, p. 553-565.
- Nelson, C. E., 2006. "Metallic Mineral Resources." In: Bundschuh, J. & Alvarado, G. (eds), Central America: *Geology, Resources and Hazards.* Chapter 32, p. 885-915. CMC Press / Balkema, Leiden, The Netherlands.
- Norton, H.K., 1930. "Why Britishers in Panama?" *The World's Work* magazine, Doubleday, Doran & Company, Inc., New York, November 1930, Vol. 59, No. 11, p. 29-32. Reproduced in St. Lawrence Waterway, Hearings before a subcommittee of the Committee on Foreign Relations, United States Senate, 72nd Congress, 2nd Session on Senate Resolution 278. United States Government Printing Office, Washington, p. 147-150.
- Oller, J., 1933. *La Industria Minera en Panama.* Biblioteca Cultura Nacional (Revista Economica Nacional), Guillermo Andreve, Panamá, Segunda Epoca, No. 19, 17 September 1933, p. 193-214.
- Oller, J., 1975. *La Industria Minera en Panama.* 2nd edition, Panamá, 22 p + 26 plates.
- Orr, J., 2018. *Scotland, Darien and the Atlantic World, 1698-1700.* Edinburgh University Press, Edinburgh, 208 p.
- Piperno, D. R., 1994. "Phytolith and charcoal evidence for prehistoric slash-and-burn agriculture in the Darien rain forest of Panama." *The Holocene*, Vol. 4, No. 3, p. 321-325.
- Prebble, J., 1968. *The Darien Disaster: A Scots Colony in the New World, 1698-1700.* Secker & Warburg, London, 1968, 366 p. Republished as *Darien: The Scottish Dream of Empire*, Birlinn Ltd., Edinburgh, 2000, 384 p.
- De Puydt, M. L., 1868. "Account of Scientific Explorations in the Isth-

mus of Darien in the years 1861 and 1865." *The Journal of the Royal Geographical Society of London*, 1868, Vol. 38, p. 69-110.

- Reclus, Armand, 1881. *Panama et Darien, Voyages d'Exploration* par Armand Reclus (1876-1898). Librairie Hachette et Cie., Paris, 422 p.
- Redwood, S. D., 2020. "The History of Mining and Mineral Exploration in Panama: From Pre-Columbian Gold Mining to Modern Copper Mining." *Boletín de la Sociedad Geológica Mexicana*, Vol. 72, No. 3, December 2020, A180720, p. 1-62. http://dx.doi.org/10.18268/BSGM2020v72n3a180720.
- Restrepo, V., 1884. *Estudio Sobre las Minas de Oro y Plata de Colombia.* Imprenta Echeverría Hermanos, Bogotá, 1st edition, 312 p.
- Restrepo, V., 1886. *A Study of the Gold & Silver Mines of Colombia.* English translation of 1884 edition by C. W. Fisher. Colombian Consulate, New York, 320 p.
- Restrepo, V., 1888. *Estudio Sobre las Minas de Oro y Plata de Colombia.* Imprenta de Silvestre y Compañia, Bogotá, 2nd edition, xvi + 334 p.
- Restrepo, V., 1937. *Estudio Sobre las Minas de Oro y Plata de Colombia.* Universidad Nacional de Colombia, Escuela Nacional de Minas Anales, 43, Medellin, 3rd edition (reprint), 314 p.
- Restrepo, V., 1952. *Estudio Sobre las Minas de Oro y Plata de Colombia.* Banco de la Republica, Bogota. 4th edition (reprint), 295 p. http://www.lablaa.org/blaavirutal/historia/minas/indice.m.
- Rodríguez Hernández, N. E., 2016. "La amenaza cuna: política de indios del gobernador Andrés Ariza en el Darien durante su primera de gobierno en el siglo XVIII." *Historia Caribe,* Vol. XI, No. 28 (January-June), p. 211-239. DOI: http://dx.doi.org/10.15648/hc.28.2016.9
- Ruiz-Williams, E., 1990. "Cana Mining District, Darien Province, Panama: Sociedad de Inversiones IXTAPA S. A.," Panama, July 1990, 7 p. Executive Summary.
- Sarasqueta Oller, G. (ed.), 2010. *Cronica sobre mis abuelos: Pasajes maritimos panamenos del siglo pasado y parte del presente.* Universal Books, Panama, 243 p.
- Sharp, Bartholomew, 1684. *The Adventures of Capt. Barth. Sharp. And Others, in the South Sea: Being a Journal of the Same.* P. A. Esq., London, 1684, p. 1-114.

• Sheridan, J.F., 1926. "Gold areas of the province of Panama." *Engineering and Mining Journal*, Vol. 122, no. 26, p. 1027-1028.

• Tippett, M.C. & Trever, P.F., 1989., "The Isthmus of Panama: A Forgotten El Dorado." *Society of Mining Engineers Preprint*, no. 89-56.

• Tower, G.W., 1964. "Cana Mines, Panama." Unpublished report for Mauricio Hochschild, January 3, 1964, 6 p. University of Wyoming, American Heritage Center, Thayer Lindslay Papers, UOW/6034/110/1.

• Williams, R.S., 1908. "Report on botanical exploration in Panama." *Journal of the New York Botanical Garden*, vol. XI, no. 105, p. 149-158.

• Wilson, F., 1981. *The Conquest of Copper Mountain. A vivid, personal account of the discovery and development of a spectacular outcrop of ore in the remote peaks of Irian Jaya, Indonesia.* Atheneum, New York, 243 p.

• Woakes, E.R., 1895. "Notes on the Espiritu Santo Mine at Cana; Its Drainage and Recovery." *Transactions of the Institution of Mining & Metallurgy*, Vol. 3, p. 285-297.

• Woakes, E. R., 1896. "Reopening of the great Espiritu Santo mine of Colombia, nearly two hundred years old." *The Engineering and Mining Journal*, New York, May 30, 1896, p. 514-515.

• Woakes, E. R., 1899. "Modern Gold-Mining in the Darien. Notes on the Reopening of the Espiritu Santo Mine at Cana." *Transactions American Institution of Mining Engineers*, Vol. 29, p. 249-280.

• Woakes, E. R., 1923. "The Darien Gold Mine, Panama." *The Mining Magazine*, November 1923, p. 270-278.

• Wozencraft, O.M., 1881. "The Wilds of the Darien." *The Californian*, Vol. 4, No. 19, August 1881, p. 80-86 (Part I); No. 20, August 1881, p. 163-171 (Part II); No. 21, September 1881, p. 280-286 (Part III).

• Wyse, L. N. B., 1877. *Canal Interocéanique, 1876-77: Rapport Sur Les Etudes De La Commission Internationale D'Exploration De L'Isthme Du Darien.* A. Chaix et Cie., Paris, 238 p. https://catalog.hathitrust.org/Record/011456448.

End Notes

1 “List of decommissioned ships of the Colombian Navy,” Wikipedia; Cuestas Gómez (1999), p. 127, 135-136.

2 Cuestas Gómez (1999), p. 127, 135-136.

3 History of the Canopy Camp Darien, http://canopytower.com/CCD-history, accessed 08-02-2022.

4 ANP, Darien Gold Mining Co. Ltd. Escritura de venta de las propiedades de Darien Gold Mining Co. In Mendez, 2004: 59-62.

5 Dolores Piperno, pers. comm., 27-02-24; Bush & Colinvaux, 1994; the latter give the lake location as 7°45’N 77°38’W which is on the ridge 5 km west of the River Tuira, and the correct coordinates are probably 7°45’00”N 77°32’20”W. No lake is visible on Google Earth or shown on the 1:100,000 topographic map but the river valley is broad and flat and probably floods in the rainy season.

6 AGI “Fortificaciones de las Castillas, de Panamá, Portobelo y Río Chagres”. Tomo XV Documento 294, 15-07-1680.

7 Cullen (1853), p. 47-48,, citing vid. t. iii. n.7 of the Archives of the Viceroyalty of Peru, Lima.

8 A Map of the British Empire in America, Sheet 18, by Henry Popple, 1733). Reproduced under Creative Commons License from the David Rumsey Map Collection, David Rumsey Map Center, Stanford Libraries.

9 Oil painting of William Dampier by Thomas Murray. ©National Portrait Gallery, London.

10 Sir John Dalrymple, *Memoirs of Great Britain and Ireland*, Vol III, 1778, Part III, Book VI, “Adventures of Paterson, and Description of Darien,” p. 121-171. Appendix, “Mr Paterson’s Last Plan for the Darien Company,” p. 154-166. The Cana gold mine is described on p. 157-159. Paterson’s bold

prospectus is discussed in Forrester's (2004, p. 295-297) biography but is ignored by most others (e.g. Prebble, 1968).

11 Image from Flickr's the Commons with no known copyright retrictions. Source Innes, J.H., 1902, New Amsterdam and its people; studies, social and topographical, of the town under Dutch and early English rule.

12 *Comentos sobre la rica i fertilisima Provincia de el Darien, en fecha i dirijidos de Santa Maria la Antigua de el Darien,* el 5 de Abril, 1774, por el Sor. Don Andres de Aríza a el Exmo. Virrei, Capitulo: De Tall de la Provincia de Sa. Maria la Antigua de Darien, arreglado al Mapa que le dirigio al Exmo. Sor. Virrei en fecha del 5 de Abril, 1774. The map was described by de Puydt (1868), Appendix III, Note on Don Andres de Aríza's Map of South Darien (1774), p. 108-109, and he presented a copy of the map to The Royal Geographical Society, London.

13 Castillero-Calvo, 2019a, p. 805, footnote 33, Letter from Governor Antonio Gill y Gonzaga to bailío Julián de Arriaga, describing the worst attacks of the Cunas, in Letter from Panamá, 21.IX.1761, AGI Panamá 260.

14 This report is titled "Comentos de la rica y fertilisima Provincia del Darien." It was published in the *Anales de la Instrucion Pública*, Number 29. English translation from Restrepo (1886).

15 1744 Plano Hidrografico y Geografico del Reino de tierra firme Nicolas Rodriguez. Archivo Historico de la Armada Juan Sebastián Elcano AMN 11-A-11. https://1drv.ms/i/c/3c81179b14f75f50/IQBQX_cUmxeBIIA8y9g KAAAAAc2SiX31H5eJ_eb4z3wIgXQ?e=JkoXO9

16 Restrepo (1884, 1886) says in a footnote that the bells and images of the church of Santa-Cruz de Cana are preserved in a chapel in the village of Molineca.

17 'Carte Générale de Darien Méridional' by Lt Lucien N.B. Wyse and Lt. A. Reclus, in Wyse (1877). The map scale is 1:250,000. The grid lines are 5' or 5 nautical miles equals 9.25 km.

18 A copy of the original Wyse 1877 map is held by the Library of Congress but is not scanned. Arauz Torres (2014: 350) reproduced a photo of the Cana area of the map.

19 Reclus (1881), 113.

20 The Gold Mines of Cana share certificates, Niko Collection and Pascual Collection.

21 TNA, BT 31/14859/24047
22 TNA, BT 31/14859/24047
23 TNA, BT 31/14859/24047.
24 TNA, BT 31/14859/24047
25 TNA, BT 31/14859/24047
26 TNA, BT 31/14859/24047.
27 TNA, BT 34/2543/24047.
28 Ernesto Restrepo Tirado, *Prologo de la Tercera Edicion*, Restrepo, 1937 and 1952, p. 19-21, biography of his father Vicente Restrepo; Vicente Restrepo Maya at EcuRed, https://www.ecured.cu/Vicente_Restrepo_Maya.
29 TNA, BT 31/14859/24047, Book List of Shareholders 14 May 1891, Vol 3.
30 Obituary, Institution of Mechanical Engineers; Grace's Guide https://www.gracesguide.co.uk/Richard_Hammersley_Heenan; Heenan-Davies, Karen (2018). "Richard Hammersley Heenan: engineer and industrial magnate" (PDF). Heenan Footprints. Accessed 09-02-2022. https://heenan.one-name.net/wp-content/uploads/2018/04/Richard-Hammersley-Heenan-profile.pdf
31 Obituary E. Woakes. *Transactions of the Institute of Mining and Metallurgy, 1947-48*, Vol. 57, p. 480-481; Northern Mine Research Society, Obituaries of Members of the Institute of Mining and Metallurgy. https://www.nmrs.org.uk/resources/obituaries-of-members/obituaries-of-members-names-beginning-with-w/ernest-russell-woakes/
32 TNA, BT 31/15178/34846.
33 TNA, BT 31/15178/34846.
34 TNA, BT 31/17974/91727.
35 Photos and notes from Peter Tristram, Australia 15-02-2022.
36 Coins of Panama.com website, David S. Ploughman. https://www.coins-of-panama.com/panamatokens/pt202-53.html Viewed 08-02-2022. David S. Plowman, 2019, Tokens of Panama / Fichas de Panama, Lulu.com, p. 59-67.
37 The steamer *Darien* appears to have been renamed *Panama* by the rebels but on recapture reverted to *Darien*.
38 "List of decommissioned ships of the Colombian Navy", Wikipedia; Cuestas Gómez (1999), p. 127, 135-136.
39 Cuestas Gómez (1999), p. 127, 135-136.
40 Photo Wikimedia, Navy History and Heritage Command.

41 Photo National Archives and Records Administration.

42 Naval History and Heritage Command, https://www.history.navy.mil/research/histories/ship-histories/danfs/b/boston-v.html.

43 Wikipedia https://en.wikipedia.org/wiki/USS_*Wyoming*_(BM-10) Navy History and Heritage Command https://www.history.navy.mil/research/histories/ship-histories/danfs/w/wyoming-ii.html (Wyoming (Monitor No. 10), 1902-1909, by Robert J. Cressman, 17 November 2021).

44 "International Boundary Study No. 62, Colombia – Panama Boundary." The Geographer, Office of the Geographer, Bureau of Intelligence and Research, US Department of State, 30 January 1966, PDF, 7 p. https://docplayer.net/211670906-International-boundary-study-colombia-panama-boundary.html, accessed 29-06-2022.

45 TNA, BT 31/15178/34846.

46 TNA, BT 31/15178/34846, Register of Directors or Managers, 29 May 1905.

47 TNA, BT 31/15178/34846.

48 TNA, BT 31/15178/34846.

49 TNA, BT 31/17974/91727.

50 TNA, BT 34/3278/91727.

51 TNA, PL 31/843. Chancery of the County Palatine of Lancaster, Manchester District, Chancery of Lane.

52 Wikipedia, *Hevea brasiliensis*, https://en.wikipedia.org/wiki/*Hevea_brasiliensis*.

53 Wikipedia, *Castilla elastica*, https://en.wikipedia.org/wiki/Castilla_elastica; Carrasquilla, L. G., 2008. *Arboles y Arbustos de Panama / Trees and Shrubs of Panama*. Editora Novo Art S.A., Panama, 478 p. C. elástica, p. 316-7. *H. brasiliensis* is not described.

54 Drabble, J. H., 1967. "The plantation rubber industry in Malaya up to 1922." *Journal of the Malaysian Branch of the Royal Asiatic Society*, Vol. 40, No. 1, p. 52-77.

55 Stephens, C. S., 2023. "From Bananas to Cacao, Abaca, Rubber and Back to Bananas." Chapter 12, *Banana Man: Peeling and Revealing, Life in the Tropics*, Book 1. Cecropia Press, Panama, p. 161-178.

56 TNA, 31/17974/91727. The Darien Gold Mining Company, Limited, Prospectus.

57 Stephens, op. cit.

58 TNA, BT 31/17974/91727.

59 TNA, BT 34/3278/91727.

60 TNA, Panama Corporation Limited, BT 31/32749/213042. Board of Trade, Companies Registration Office, Files of Dissolved Companies.

61 TNA, Panama Corporation Limited, Prospectus 15 April 1926, BT 31/32749/213042.

62 Mackie, C.A., 2013. *A Directory of British Diplomats*. Updated 11 February 2016, 1285 p., online.

63 TNA, Panama Corporation Limited, Prospectus 15 April 1926, BT 31/32749/213042.

64 Hull, 1940; TNA, Panama Corporation Limited, BT 31/32749/213042.

65 TNA, Panama Corporation Limited, BT 31/32749/213042. The British company was wound up in 1948.

66 Richard Oglesby Marsh, Public domain. Harris & Ewing, photographer - Library of Congress Catalog: https://lccn.loc.gov/2016893798.

67 Reproduced with the permission of Matthias Schmitt, HWPH Historisches Wertpapierhaus AG, Germany, https://www.hwph.de.

68 Panama Land Concessions: Appeal to the Monroe Doctrine. The West Australian, 2 August 1926, p. 15. https://trove.nla.gov.au/newspaper/article/31939991/2781621. Accessed 15-02-2022.

69 Richard Oglesby Marsh, Wikipedia, https://en.wikipedia.org/wiki/Richard_Oglesby_Marsh. Accessed 15-02-2022.

70 Obituary V.F.S. Low, *Transactions of the Institute of Mining and Metallurgy, 1956-57*, Vol. 66, p.596; Northern Mine Research Society, Obituaries of Members of the Institute of Mining and Metallurgy. https://www.nmrs.org.uk/resources/obituaries-of-members/obituaries-of-members-names-beginning-with-l/vernon-forster-stanley-low/

71 Biographical details of Alves from RootsChat genealogical site, Duncan Elliott Alves and wife Tunbridge Wells forum, https://www.rootschat.com/forum/index.php?topic=732870.0, accessed 03-12-22; '1914: The Generous Businessman'. Tunbridge Tales, 2 March 2015. https://tunbridgetales. com/2015/03/02/1914 -the-generous-businessman/. Accessed 21-05-2020. Wikipedia, Bryn Bras Castle, https://en.wikipedia.org/wiki/Bryn_Bras_Castle, accessed 03-12-22.

72 The Nottingham Evening Post, 20 December 1910, p. 6, 'Death of a New

Zealand Colliery Proprietor', posted at Alves Family of Victoria forum https://www.rootschat.com/forum/index.php?topic=732932.18, accessed 03-12-22.

73 Bryn Bras Castle Grade II Listed Building, report, http://www.coflein.gov.uk/pdf/CPG041/, accessed 03-12-22.

74 Large image unknown artist © The Caernarfon Royal Town Council Collection accessed at ArtUK.org. Small image from the Kent & *Sussex Courier*, 6 th May 1910, p. 3. Both accessed at Tunbridge Tales, op. cit., with thanks to Anne Carwardine.

75 Hugh Marriott obituary, Trans. I.M.M., Vol. 59, 1949-50, https://www.nmrs.org.uk/resources/obituaries-of-members.

76 Walter Broadbridge obituary, Trans I.M.M, Vol. 56., 1946-47, p. 609-610.

77 UOW/6034/110/1 Letter from George W. Tower, New York to Dr. Mauricio Hochschild, Paris January 27, 1964.

78 'Louis Martinz; Former Panamanian Spokesman, 54', *New York Times*, 30 December 1994; G. Ellis, pers. com., 2007, a geologist who worked on the project at this time.

79 UOW/6034/110/1 Letter from George W. Tower to Dr. Mauricio Hochschild, January 3, 1964; Abstract of letter from William H. Breeding, dated January 10, 1964 [to George W. Tower]; Tower, 1964.

80 Tower, 1963.

81 UOW/6034/110/1 Letter from George W. Tower to Dr. Mauricio Hochschild, January 3, 1964; Tower, 1964.

82 UOW/6034/110/1 Letter from George W. Tower to Dr. Mauricio Hochschild, January 3, 1964; Tower, 1963.

83 Fundacion Gabriel Lewis Galindo, Don Gabriel Lewis Galindo, http://fglg.org.pa/don-gabriel-lewis-galindo/, accessed 10-02-2022. See biography by De La Guardia de Corró (2009).

84 Raymond Jannas, ex-Gold Fields geologist who worked on the project, pers.com.

85 Gwendy E. M. Hall, Obituary for John Alexander Hansuld (1931-2019), Elements magazine (Association of Applied Geochemists), October 2020, Vol. 16, No. 5, p. 350; Canadian Mining Hall of Fame, 2012 Inductee, John A. Hansuld, https://www.mininghalloffame.ca/john-a-hansuld.

The Author

DR. STEWART D. REDWOOD, BSc (Hons), PhD, FGS, FIMMM, QMR, FSA Scot is a globe-trotting professional geologist who works as an independent consultant in the minerals exploration business. Originally from Scotland, he has lived in Panama for more than 30 years. He has a first class honors degree in geology from the University of Glasgow and a doctorate in geology from the University of Aberdeen for research on the gold, silver, copper and tin deposits of Bolivia. He served in the Royal Navy reserves, explored for oil in the Norwegian Arctic, and worked for the British Geological Survey before moving to South America in 1989 as the Director of Exploration for a Bolivian consulting firm. He has been involved in the discovery of several mineral deposits of gold, silver, copper and zinc in Bolivia, Peru, Colombia, Dominican Republic, Turkey, Sweden and other countries that have increased the GDP of those countries. He has written numerous scientific papers and technical reports. He has an interest in mining history and has written extensively about Panama. He is the publisher of an e-magazine, ***Tales of Old Panama and Castilla de Oro***, since 2020. The present book is about the most famous gold mine in Panama. He lives with his Bolivian wife Maite and several cats in Panama City and La Paz.

Thank you for buying this book. I hope you enjoyed reading it as much as I enjoyed writing it. If you can, please visit wherever you bought the book and leave a review, and recommend it to your friends.

You can write to me at stewart@sredwood.com.

S.D.R.

Cecropia Press
cecropiapress.com

Made in United States
Orlando, FL
04 November 2026

90200956R00141